P9-CFJ-124

THE SNAKE CHARMER

Also by Jamie James

NONFICTION

The Music of the Spheres

Other Origins
(with John Olsen and Russell Ciochon)

Eccentrics: A Study of Sanity and Strangeness
(with David Weeks)

FICTION

Andrew & Joey: A Tale of Bali

THE

SNAKE CHARMER

A Life and Death in Pursuit of Knowledge

JAMIE JAMES

HYPERION
NEW YORK

 For information address Hyperion, 77 West 66th Street, New York, New York 10023-6298.

Library of Congress Cataloging-in-Publication Data
James, Jamie.
The snake charmer : a life and death in pursuit of knowledge / Jamie James.
p. cm.
ISBN-13: 978-1-4013-0213-9
1. Slowinski, Joseph. 2. Herpetologists—United States—Biography. 1. Title.
QL31.J56J36 2008
597.9092—dc22
[B] 2007048987

Hyperion books are available for special promotions, premiums, or corporate training. For details contact Michael Rentas, Proprietary Markets, Hyperion, 77 West 66th Street, 12th floor, New York, New York 10023, or call 212-456-0133.

FIRST EDITION

10 9 8 7 6 5 4 3 2 1

Designed by Jessica Shatan Heslin/Studio Shatan, Inc.
Illustrations and endpaper map by Bruce Granquist
Photo on title page spread © Ashleigh B. Smythe

For Davien Littlefield,

always a step ahead

Contents

Many-Banded Krait

No snake kills with more ruthless efficiency than the many-banded krait, which dwells in the jungles of India and Southeast Asia. Drop for drop, its venom is the deadliest of any land serpent's, apart from a few rare species found only in the outback of Australia. One bite of the krait carries enough concentrated toxin to kill two dozen grown men.

American soldiers during the war in Vietnam called it the "two-step snake," in the belief that its venom is so lethal that if it bites you, you will fall dead after taking just two steps. That's an exaggeration, but the bite of the many-banded krait is astonishingly potent. The venom is a neurotoxin, which means that it disables the victim's nervous system—like yanking an electrical plug out of the socket. Death comes when neurotransmission ceases: With no instructions to breathe, the muscles of the diaphragm are stilled, and the victim asphyxiates.

Usually the victim of the many-banded krait is another snake; because the species is cannibalistic, it might even be another krait. Yet it never seeks larger prey. The last thing it wants is a brush with a human being; the snake is far more likely to end up dead from the encounter than the person is. But the many-banded krait is built to kill, and if it is threatened, it can only do what it is programmed to do: It bites.

Most people go to great lengths to avoid meeting a many-banded krait, but in 2001, biologist Joe Slowinski traveled from San Francisco, where he was a curator of herpetology at the California Academy of Sciences, to Upper Burma, expressly to look for them—and all the other reptiles and amphibians he could find. Joe was one of the leading experts in the world on the venomous snakes of Asia, and his latest expedition was the most ambitious scientific mission in Burma's history: He was leading fifteen naturalists and more than a hundred Burmese support staff into wilderness terrain that was scarcely known to science.

It might seem logical to deduce from the wide availability and superfine detail of satellite mapping that the wilderness of Earth has now been fully charted, all its secrets exposed. Yet that is far from the case: There are still vast tracts of the planet that remain almost unexplored. Burma, which many people now call Myanmar, the Southeast Asian country west of Thailand, is one of the most poorly studied places in the world. Neglected by British scientists in colonial times, ripped apart by civil war in the postwar period, and brutally plundered by its own corrupt military regime since a coup in 1988, Burma hasn't had time for science. In his memoir *Burma's Icy Mountains*, the British botanist Frank Kingdon-Ward, one of the few foreign explorers to venture there, observed after an expedition in 1937, "North Burma is an excellent example of a country which is surveyed but not explored."

Since Kingdon-Ward's day, much of the rest of the world has been studied intensively, but Burma remains terra virtually incognita—a potential treasure house for an enterprising biologist. When a former student of Joe Slowinski's discovered a new species of nonvenomous snake in Louisiana a few months before Joe's expedition, it was the first new serpent species to be identified in North America in more than half a century. Yet the wildlife of Burma remains so little known that Joe and his small band of American and Burmese colleagues discovered new species on virtually every expedition and projected that their research would eventually yield more than fifty previously unknown reptiles and amphibians—perhaps as many as one hundred and fifty. Here, in the foothills of the Himalayas, the most remote region he had yet studied, Joe expected to discover many new species.

Joe Slowinski and his team—botanists, ichthyologists, and ornithologists, experts on insects and mammals—were foot soldiers in Darwin's army. Modern biologists, armed with sophisticated technology, are continuing the great intellectual adventure of compiling a comprehensive census of life on Earth, an audacious enterprise launched in the Enlightenment and given its theoretical foundation by Darwin himself, and carrying it forward into the twenty-first century.

It was late summer, the stormy end of the monsoon. No sensible person who wanted to come to Putao District, Kachin State, in uppermost Burma, would contemplate going there during the rainy season. But frogs love water, and where there are frogs, there are snakes to eat them. Joe had expected muddy trails, bad food, and squalid campsites; as far as he was concerned, that was all part of the fun. Yet this expedition set a new standard of misery. The rain had poured almost incessantly since they set out from Putao, the small town that served as the administrative capital of this isolated region. The trail was a deep river of fine, clinging clay mud. Malarial mosquitoes and sharp-biting sandflies swirled in tormenting clouds; legions of thirsty leeches lurked in every dank, dark recess.

Most dispiriting of all, the snake collecting had been poor. Yes, rain brings out snakes, but not when it's as heavy as this. How can you find snakes to catch when the rain sheets down so densely that you can scarcely make out the back of the hiker ten feet ahead of you?

On September 9, the expedition reached a village called Rat Baw, by far the most advanced settlement the group had seen since leaving Putao. It even had a street paved with cobblestones and a frame schoolhouse with a tin roof—rare luxuries in a region where most people shared smoky bamboo huts with their livestock. The scientists made their camp in the school, the first dry berth they had had in many days. Their luck finally seemed to be changing: The afternoon they arrived the sun came out—and so did the snakes. Joe and the other herpetologists immediately headed into the forest to search for reptiles.

Nobody alive knew more about kraits than Joe Slowinski did. One thing he knew was how tricky they can be to identify, for there are nonvenomous snakes that mimic them with amazing accuracy. Joe had recently discovered one such species and named it himself. *Lycodon zawi*, a wolf snake, imitates the krait's alternating black and white bands so closely that it's difficult to make a definite identification from a few feet away.

Another species, *Dinodon septentrionalis*, is a dead ringer for the krait. The nonvenomous Dinodon mimics the many-banded krait so uncannily well that even an experienced herpetologist might need a magnifying glass to tell them apart. The only way to be sure is to examine the snake's head, to see whether what is known as the loreal scale is present, just below the eye: a fleck scarcely bigger than a snowflake, a tiny pentagon of horny skin that tells the difference between a harmless, handsomely banded Dinodon and the lethal krait.

This classic example of mimicry evolved millions of years ago, when predators avoided snakes that resembled the venomous krait, giving them an edge for survival. Nature didn't give the Dinodon venom, but it got the next best thing: a resemblance to the deadly many-banded krait, which makes snake-eating predators shy off.

More interesting to Joe, however, were the snakes that he and the other herpetologists were still carefully logging as *Bungarus multicinctus*; the many-banded krait (a name coined in 1861 by Edward Blyth, a British zoologist in Calcutta) probably belonged to other species. In 2001, the consensus was that the scientific name *Bungarus multicinctus* comprised several different species of krait, which were still waiting to be collected and identified by a skilled herpetologist. Joe Slowinski thought he was just the man for the job.

One of the most telling documents of the expedition is a photograph of Joe standing in a patch of sunlight, wrangling a many-banded krait.

The ostensible purpose of the photograph was to serve as an archival record of the specimen, but the shot also powerfully conveys Joe's confident mastery of the situation. The children of the village came out to gawk as the pale-skinned visitor, taller and stockier than the malnourished men who lived there, free-handled the dangerous serpent they called *ngan taw kyar* ("royal tiger snake") with cool bravado. Their parents hung back, looking over a twig fence, watching in astonishment as the krait twisted and wriggled in the stranger's expert hands. Joe was a brilliant biologist in his prime, but he took a visceral, almost rapturous delight in handling snakes. This pleasure shines through in the photograph, captured in a glorious interlude of sun: The light gleams on Joe's mop of tawny hair and goatee, glinting penny-bright.

At thirty-eight, Joe's life was riding a dizzying upward arc. Four years earlier, he'd been toiling in a dead-end job as a lecturer at a small university in Louisiana; now he was among the most respected and influential herpetologists in America. A few days before Joe left San Francisco for the expedition, the National Science Foundation had awarded him a grant for $2.4 million—the largest public research grant the California Academy of Sciences had received in its 148-year history. By year's end, he would take over the chairmanship of the herpetology department at the Academy, the premier natural-history museum in the American West. And Joe was in love—unexpectedly and passionately in love—with a beautiful woman he'd recently met in San Francisco. With the toad middle age squatting in the middle of his path, leering, he was thinking of settling down. In Burma, Joe dreamed of Sandy.

That evening, the group received bad news. Runners from the next village on the expedition's route arrived in Rat Baw, reporting that mudslides had closed roads, and floods had washed out bridges on the trail ahead. Joe had planned to press on soon into the Himalayas, even more remote terrain; now it appeared that the journey onward would almost certainly have to be canceled.

Yet the next morning Joe rose early and bounded out of his sleeping bag into the gray mist of first light. He found his Burmese assistant already hard at work on the porch, attempting to bring some order to the specimens collected the day before. The snakes were in cloth bags, lined up on the assistant's worktable. Every now and then, one of the bags would wiggle and thump.

Joe asked about a particular specimen, and the assistant handed it to him, saying that he thought it was a Dinodon.

Although it was still too dark to see well, Joe absentmindedly thrust his right hand into the sack to extract the specimen and have a look. Immediately, he winced with pain and yanked out his hand. A tiny black-and-white banded snake, less than ten inches long, was dangling limply from his middle finger, its fangs still sunk into his flesh.

Joe looked at it in quiet horror. Without the hope of a doubt, he said, "That's a fucking krait."

The Snake Charmer

Black Rat Snake

Elaphe obsoleta

The black rat snake, which inhabits the eastern United States from New England to Oklahoma, is a large, powerful constrictor and excellent climber. The animals' shed skins are often found draped in the rafters of barns and abandoned buildings. They can climb trees up to forty feet high; with their cryptic black coloration, they become invisible against tree bark or at rest on the forest floor. The species is nonvenomous, and feeds on mice, chipmunks, voles, shrews—even squirrels and baby birds. Black rat snakes are sociable hibernators, sharing their rocky crevice retreats with many other snake species, including timber rattlesnakes, racers, and bull snakes. If threatened, they may vibrate their tails, producing a rattling sound, which sometimes causes the species to be killed by frightened people who mistake them for rattlesnakes.

Joe Slowinski's connection with the animal world was always intense, immediate, and electric. And he discovered it all by himself.

It began in the summer of 1968, in Wisconsin. Joe was five years old. His father, Ronald Slowinski, an artist and professor of painting at the Kansas City Art Institute, was appointed to a summer residency at the Peninsula Art School in a small town called Fish Creek, near Green Bay. The Slowinskis—Ron and his wife, Martha, Joe, and his baby sister, Rachel—were housed in a cabin made of hand-hewn logs, which had aged to a silvery russet.

Bucolic Door County, Wisconsin, was the perfect place for a curious boy to start his exploration of the natural world. One day, Joe found a butterfly chrysalis in the milkweed near the cabin. It was an exquisite thing, a pale aquamarine lozenge pranked with spots that gleamed like drops of pure gold. No one was sure exactly what it was, but Martha found a glass jar for it, and Joe punched airholes in the lid. A couple of days later the chrysalis turned dark, and the next morning, while Joe and his mother were watching, a monarch butterfly emerged. When Joe opened the jar, the butterfly crawled to freedom, shook out its delicate wings, and flew away. Martha said, "It was enough to make a naturalist of anyone."

It was an idyllic summer. During the days, Joe and Rachel stayed at a day-care center in a farmhouse nearby, watched over by some young mothers from the neighborhood, while their parents made art. Ron painted his big geometric abstractions in the nobly proportioned stone barn provided for him as a studio. Martha, a figurative painter, worked there too, but she also loved to draw outdoors, sketching the gnarled apple trees in the abandoned orchard that surrounded the barn. Their bare, twisted branches reminded her of Van Gogh.

One sunny morning, this tranquil scene was shattered by hysterical screams emanating from the day-care center. Ron, in his studio, and Martha, sketching outside, both heard them. Dropping paintbrush and charcoal, the young parents ran toward their children on converging paths through the tall weeds that fringed the orchard. They found Joe surrounded by a knot of distraught adults, enjoying himself enormously as he brandished a writhing black snake over his head—a snake longer than he was, and as thick as his own little arm.

Perhaps some adult had told Joe that big black snakes like that weren't dangerous. More likely, it never occurred to him to be afraid: It just wasn't in his nature. He was entranced by the strength and strange beauty of the animal, a living creature with a will of its own, now subjugated to him—and as happy and pleased with himself as a little boy can be who has given the grown-ups a good scare.

By the time Ron and Martha got there, some local men had arrived who recognized that it was a harmless rat snake. Joe was ordered to turn the snake loose and scolded for his naughtiness, but halfheartedly: Everyone was relieved that no harm had been done.

Throughout his life, Joe never met an animal he feared. His father would later say, "Joe always thought of himself as a critter, not a person."

The black rat snake is one of the nearly three thousand species belonging to the suborder Serpentes. Although the rational system of classifying earthly life forms, known as taxonomy, is far more complex now than when its elements were codified in the eighteenth century, nearly all snakes are still classified as belonging to one of three families.

Most species usually described as nonvenomous, like the black rat snake, are grouped together as the Colubridae. Joe Slowinski was coauthor of a paper published in 2005 that described this taxon, or category of related life forms, with a concision unusual in the scientific literature: "The family Colubridae is the most diverse, widespread, and species-rich family within all of Serpentes, occupying all continents except Antarctica and consisting of greater than 1,800 species." Thus the colubrids actually account for a majority of all known snakes. *Dinodon septentrionalis* is a colubrid, as are common garter snakes.

Most dangerously venomous snakes, on the other hand, fall into one of two families: the Viperidae, comprising vipers such as copperheads, rattlesnakes, and the deadly Russell's viper of Asia; or the Elapidae, distinguished by having permanently erect fangs fixed in the front of the jaw, which include the cobras and kraits of Asia, most of the snakes of Australia, and the coral snakes of the Americas.

The taxonomy of snakes is more complicated than that, of course.

This scheme omits several small, fascinating families, such as the primitive, subterranean blind snakes, no bigger than worms; rare, transitional forms such as the Indonesian dwarf pipe snakes; the so-called basal snakes, the prodigiously robust pythons and boas; the venomous, burrowing Atractaspididae; alluring, iridescent sunbeam snakes; and other taxa that do not have a high degree of speciation (number of species) or wide distribution. Many scientists nowadays are abandoning the cozy cupboards of class, order, family, and so forth, created in the 1700s by the Swedish botanist Carolus Linnaeus; they prefer to think in terms of lines of descent from common ancestors, called clades, which are based upon analysis of species' shared DNA.

Like most people, herpetologists typically take a greater interest in snakes that pose a threat to human life than to those that don't. Bryan Fry, an expert on snake venom at the University of Melbourne (with whom Joe had planned a rendezvous in Singapore at the conclusion of his expedition to Upper Burma in 2001), describes the Colubridae as "a bit of a fiction—a dumping ground for any snakes that aren't vipers or elapids." In a paper published after his expedition to Upper Burma, Joe and his coauthors analyzed the DNA of a hundred colubrid species, and found that some of them are actually closer to the elapids, the deadliest snakes on Earth, than they are to other colubrids.

Since the family was first designated in the late nineteenth century, the Colubridae have always been described as nonvenomous; yet Fry has discovered that the bite of nearly all of them, even humble racers and garter snakes, is poisonous. "There's just enough toxin to stun or slow down the prey," said Fry. "That doesn't mean that they're dangerous to humans: How much does it take to slow down a gecko or a vole? Venom is just one more tool in the snake's hunting arsenal."

The widespread assumption that all colubrids are nonvenomous had begun to lose credence by the end of the twentieth century. In 2000, Joe Slowinski wrote an article for *California Wild*, the members' magazine published by the California Academy of Sciences, in which he estimated that "at least 25 percent" of colubrids are venomous. In 2006, Bryan Fry said, "Now it looks like it's closer to a hundred percent. It's one of the great oversights in science." That's an exaggeration, but the discovery that the bite of common garter snakes is toxic has overturned the reassuring advice given to generations of children by well-meaning scoutmasters and schoolteachers.

Joseph Bruno Slowinski was born in New York City on November 15, 1962; like most children in Manhattan, he had his first memorable encounters with animals at the Central Park Zoo. He was a beautiful boy, with flossy golden hair and lucid blue eyes, who was quite capable of attracting attention without brandishing a black rat snake. When he was two years old, Joe and his father were wandering through the park after a visit to the zoo when they passed Elizabeth Taylor and Richard Burton on a footpath near Fifth Avenue. It was 1964, the year of Taylor and Burton's scandalous affair on the set of *Cleopatra*. Liz Taylor, the most famous movie star in the world, walked up to Joe, patted him on the head, and said, "What a cute little boy."

The Slowinskis lived in a walk-up loft in the neighborhood now known as SoHo, when it was still a grimy industrial district. Ron Slowinski was supporting himself as many young artists in New York did, by working at a museum. The conservation department at the Guggenheim paid him eighty dollars a week, nearly starvation wages for the young family, but both parents were resolutely committed to their artistic vocation. They made do.

Ron had grown up in a big Polish family on the West Side of Chicago—"stinking, rotten, obscenely poor," he would later say. His father was a chronic alcoholic, a kindhearted weakling who died when Ron was twelve. It was a close-knit family, benignly ruled by Ron's maternal grandmother, who insisted on two cardinal virtues: love of family and obedience to the Roman Catholic Church.

Martha Crow had grown up an independent-minded country girl with seven brothers and sisters, in Loudonville, Ohio, in the hills west of Akron. She and her twin, Mary, had spent much of their childhood outdoors. Martha was always looking for snakes, and sometimes found black rat snakes—perfect for terrorizing her sister. Their father had been adamant that his children attend and graduate from college; eventually, they all did.

Ron met Martha in 1955, when he was in the U.S. Army, stationed at Belle Isle, Michigan. They began dating six years later, when they met again in New York. Ron Slowinski was a devout Catholic; when Joe was born, he chose his middle name, Bruno, in tribute to the saint who had founded the Carthusians, the Church's most ascetic order. Martha was an atheist, but she acceded to Ron's request that they marry in the Church, and she even attended a Church-sponsored marriage class to prepare for the rite, which moved Ron and earned his respect. Fourteen months after Joe, his sister, Rachel, was born.

The children's early life was a peripatetic one. Soon after Joe's brush with celebrity in Central Park, Ron Slowinski got a real job: Indiana University, in Bloomington, hired him to teach painting, to replace the renowned still-life painter William Bailey. One year later came an invitation to join the faculty of the Kansas City Art Institute, one of the nation's best art schools. Before he got the job, Ron said, he couldn't have found Kansas City on a map. But he went to Missouri to have a look, and took an instant liking to the gentle, verdant landscape and the human scale of the city's urban design, which is based upon the enlightened principles of the late-nineteenth-century City Beautiful movement. Ron and Martha rented a faux half-timbered, two-story Tudor house in the Crestwood district, on Cherry Street. The family's quality of life had improved decisively.

After just two years at the Art Institute, Ron Slowinski took a sabbatical: Prodded by Martha, he had applied for and won a Fulbright fellowship to go to Japan to paint. Shortly after returning from Wisconsin, the family was off to Kyoto for the year. Ron and Martha resolved to live an authentic, traditional Japanese life during their time there. They often dressed in Japanese style at home; they ate Japanese food; they lived in a traditional house with tatami mats and paper windows, and a room dedicated to the tea ceremony. Like most Japanese houses, it was icy cold in the winter; a portable kerosene stove was the only heat source. The children were immersed in a way of life far from the lawns of Kansas City.

Rachel idolized her brother when they were small. During that year in Kyoto, with no other Western children to play with, they became best friends. The Slowinskis' garden gate opened onto the banks of the Takano River, which winds through the ancient city. In the summer, the children spent the long evenings stomping around on the gravelly banks of the river with butterfly nets and fishing gear, in search of adventure—which usually meant looking for wildlife.

Joe didn't simply want to observe the animals he found: He wanted to possess them. A rusticated rock wall ran along their street, which had a burgeoning population of lizards living in its cracks. Joe developed his own way of catching them: He would poke a pencil into a crack until a lizard bit the point, then yank the creature out and grab it with his other hand. One day Joe saw a big snake crawling in the garden behind the house. Remembering the black rat snake in Wisconsin, he leaped about frantically, crying, "I'm going to catch that snake! I'm going to make it my pet!" Yet this time, the snake was too fast for him.

That same year, on a family excursion to one of Kyoto's historic gardens, Joe was instantly attracted to the koi—burnt-orange carp with inky black splotches—that were swimming circles in a decorative pond. He told Rachel over and over, "I'm going to get me one of those koi." Although she was just five herself, Rachel knew her brother well enough to see how it would end, but he refused to listen to her warnings. Soon Joe himself was in the pond, and the monks were running around flapping their kimono sleeves and shouting. "He always had that burning curiosity," said Rachel. "The attitude 'I *will* make this happen.'"

Upon their return to Kansas City, the Slowinskis moved into a new house around the corner, on East Fifty-fifth Street. There was a small park behind the house, a pleasant grass oval shaded by a grove of trees sheltering a spring, which according to local legend had watered horses during the Civil War. In those days, Kansas City ended at 110th Street. The woods and plains of rural Missouri and neighboring Kansas were just a short distance away. Soon Joe was bringing home everything animal, vegetable, or mineral that excited his interest.

His collecting habit began early, with a stamp album his grandmother gave him for Christmas. Poring over foreign stamps prompted his first daydreams about traveling to exotic, faraway places. His favorite stamp was triangular, with the exciting image of a leaping tiger, from Burma. Soon he moved on to rocks and minerals. His father found a handsome oak sorting cabinet, a relic of a small-town post office, at a junk store in Kansas City, and bought it for Joe's collection. Joe carefully classified and labeled his rocks, and eventually filled all the pigeonholes in the cupboard.

Ron Slowinski was also a collector. He later built a museum-quality collection of Indian art and artifacts, and inculcated in his son a deep respect for American Indian cultures. One of the earliest proofs of Joe's prodigious talent as a finder of things of scientific value came when he hunted Indian arrowheads at his uncle David's house in rural Ohio, near Upper Sandusky. He was seven years old. Noticing that a field behind his uncle's house had just been plowed, Joe asked if he could go there to look for arrowheads. The grown-ups smiled at his childish fancy and said yes. A few hours later, Joe returned with an Indian arrowhead in his hand.

When Joe was eleven, Martha saw an announcement in the newspaper about a geology club at the University of Missouri–Kansas City, just a few blocks from home. The Heart of America Geology Club met in an old-fashioned science classroom with slate-topped lab tables, where the members would display their latest finds. Joe never missed a meeting; if his mother couldn't go, he would walk. His enthusiasm and intellectual precocity soon endeared him to the members, who were mostly geology students from the university or elderly amateurs.

Soon Joe took to arriving before the meetings began, in the hope that a friendly graduate student would invite him down to the geology lab to explore. Joe was fascinated by the place, with its fabulous chrome instruments capable of penetrating the mysteries of the geodes, gems, and fossils that lined the lab's wooden cases—the adult-world model for his post-office cupboard at home. He felt an instant sense of belonging in this cool, clean zone devoted to knowledge, which soon evolved into an ambition to be a scientist when he grew up.

At the Heart of America Geology Club, Joe met the first of his mentors, a retired physician from Shawnee Mission, Kansas, named Richard Sutton. Dr. Sutton encouraged the boy's curiosity, and gave him adult reassurance that he could make his dream of a career in science a reality. Dr. Sutton had a fine collection of ammonites—fossils of an extinct order of mollusks that had a chambered shell like that of the nautilus—which Joe studied under his guidance at club meetings.

Dr. Sutton gave Joe a few bones he had brought home from safaris in Africa in the 1920s, as well as the pride of Joe's collection: a battered human skeleton called Long John Silver, because it was missing its right leg. At Halloween, Joe daubed Long John's bones with fluorescent paint and hung the skeleton from a noose in a haunted house he created in the family basement. Joe's room was becoming a hall of wonders. Rachel would later describe her brother at this age as a boy P. T. Barnum.

In 1973, Joe's family rented a historic adobe house in Taos, New Mexico, for the summer, the first of many sojourns at the high-desert mesa for the children. On one visit, Ron took Joe and Rachel to see the Hopi snake dance, one of the most powerful and mysterious North American Indian rituals. (Access to the ceremony was already beginning to be restricted; these days most Hopi villages have returned to the old policy of excluding outsiders from attending the ceremony altogether.)

An elite brotherhood of Hopi priests devoted to snake worship enacts this spectacle once a year, always in August. The exact timing of the ceremony is determined by esoteric signs; according to one legend, the date is chosen when the sun casts a shadow from a certain rock in a certain way. The ritual is a prayer for rain, intended to win the snakes' sympathy.

Before the event begins, the priests of the clan are sent into the desert in the four cardinal directions to gather snakes. They must catch every snake they see. Then they bring the whole collection to a holy bower of cottonwood branches, the priests' sanctum, where they commune with the snakes in secrecy for four days before the dance. At the ceremony's conclusion, the snakes are dispatched to the gods of the underworld, to petition them for rain on behalf of the human race.

After he left office, President Theodore Roosevelt visited the "Hopi mesa towns, perched in such boldly picturesque fashion on high, sheer-walled rock ridges," and was given the rare privilege of visiting the Hopi priests before a performance of the snake dance. In his memoir *A Book-Lover's Holidays in the Open*, published in 1916, Roosevelt gives an intimate look at the ritual, which had probably changed little by the time Joe Slowinski witnessed it sixty years later:

> As a former great chief at Washington I was admitted to the sacred room, or one-roomed house, the kiva, in which the chosen snake priests had for a fortnight been getting ready for the sacred dance. . . . Entrance to the house, which was sunk in the rock, was through a hole in the roof, down a ladder across whose top hung a cord from which fluttered three eagle plumes and dangled three small animal skins. Below was a room perhaps fifteen feet by twenty-five. One end of it, occupying perhaps a third of its length, was raised a foot above the rest, and the ladder led down to this raised part. Against the rear wall of this raised part or dais lay thirty-odd rattlesnakes, most of them in a twined heap in one corner, but a dozen by themselves scattered along the wall. There was also a pot containing several striped ribbon-snakes, too lively to be left at large. Eight or ten priests, some old, some young, sat on the floor in the lower and larger two-thirds of the room, and greeted me with grave courtesy. . . .
>
> Some of the priests were smoking—for pleasure, not ceremonially—and they were working at parts of the ceremonial dress. One had a cast rattlesnake skin which he was chewing, to limber it up, just as Sioux squaws used to chew buckskin. Another was fixing a leather apron with pendent thongs; he stood up and tried it on. All were scantily clad, in breech-clouts or short kilts or loin flaps; their naked, copper-red bodies, lithe and sinewy, shone, and each had been splashed in two or three places with a blotch or streak of white paint.

Ron led Joe and Rachel to the roof of a building overlooking the plaza to watch the snake dance. At the ceremony they attended, there were about fifty venomous and nonvenomous snakes: many sidewinders and rattlesnakes, as well as rat snakes, bull snakes, and garter snakes (Roosevelt's "ribbon-snakes"). One of the priests approached the cottonwood bower, where he picked a snake from a large ceramic pot. The priest bit the snake about four inches behind the head, letting its tail dangle down his bare torso.

He danced with another celebrant called the hugger, who carried a feathered wand to subdue the snakes if they started to become restless. The two danced a couple of circuits around the plaza and then let the snake go. They repeated the dance with every snake in the bower. Another class of priests, the snake gatherers, were charged with keeping an eye on the snakes that had been let loose in the plaza, and to pick them up if they looked as though they were about to jump into the audience. By the end of the dance Joe attended, the snake gatherer had half a dozen snakes in each hand, gripping them below the jaws.

At one point, as a snake gatherer was picking up a wayward sidewinder, the venomous viper snapped its head around and bit the priest in the thigh. He let go of the snake, and it flopped down against his leg, still hanging on by the fangs sunk deep in his flesh. Gently, calmly, the priest removed the snake and carried on with the dance. This scene took place just below the Slowinskis' rooftop perch. Joe, his father said, "was freaking out, staring at the spectacle in total fascination."

At the conclusion of the ceremony, the priests drew a big circle with cornmeal in the middle of the plaza and threw all the snakes into the center. There was a moment of chaos as the serpents writhed in a mass. Then they dashed for the corners of the plaza and shot back into the desert.

Science began to pay serious attention to the subject of venomous colubrids after the death of Karl Schmidt, the curator of the reptile collection at the Field Museum of Natural History in Chicago, who was regarded for many years as the dean of American herpetologists.

Schmidt made his name by cataloguing four thousand specimens collected in the Congo by James Paul Chapin and Herbert Lang, who led a heroic expedition covering fifteen thousand miles of jungle—mostly on foot and by canoe—from 1909 to 1915. Schmidt founded the collection of amphibians and reptiles at the Field Museum in 1922, making it into one of the nation's leading research centers for herpetology.

His exceptional career ended in 1957, at the age of sixty-seven, when the Lincoln Park Zoo sent over a snake to be identified. With his wide experience in African reptiles, Schmidt immediately identified it as a smallish boomslang (*Dispholidus typus*), a bug-eyed colubrid from the southern part of the continent. While he was handling the snake, it nicked his thumb with a single fang. He felt a bit nauseated, but he wasn't concerned and didn't seek medical attention. The next morning he felt fine, but by mid-afternoon he was dead of a brain hemorrhage and respiratory collapse.

A similar destiny awaited Robert Mertens, a German herpetologist whose career paralleled in Europe the phenomenal success of Karl Schmidt's in America. In 1919, at the age of twenty-five, Mertens was hired as a curator by the Senckenberg Museum in Frankfurt. Over the next forty years, working virtually alone except for the help of his wife, he made the Senckenberg into one of the world's finest natural history museums. Mertens was indefatigable in his efforts to expand the collection. During World War II, he persuaded German soldiers to send him snake specimens from occupied countries by the efficient post of the Wehrmacht. He died at home at the age of eighty-one, after being bitten by a pet African twig snake he was hand-feeding. Mertens lingered for eighteen agonizing days, during which he kept a journal describing his symptoms. It concluded shortly before his death on a note of black humor: "A singularly appropriate end for a herpetologist."

Yet by the close of the twentieth century, no academic herpetologist had ever died of a snakebite in the field.

Boa Constrictor

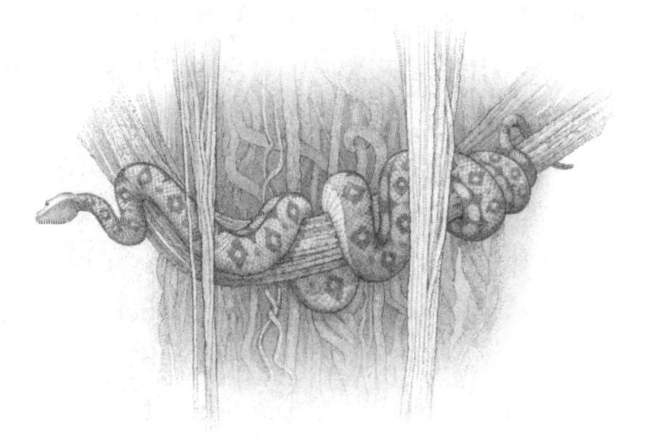

The boa constrictor is an aggressive hunter in the rain forests of Central and South America that famously kills its prey not with venom but by coiling its robust body around the victim and squeezing the life out of it. Once the prey is unconscious or dead, the boa swallows it whole. The boa constrictor is a nocturnal hunter, which uses thermal-sensitive facial scales to locate prey. The snake feeds on lizards, birds, and small mammals such as opossums, mongooses, rats, and squirrels; occasionally the boa consumes animals as large as the ocelot and the white-tailed deer, which outweigh it. The snake's preferred prey are bats, which it catches by hanging from the branches of trees or the mouths of caves and snatching the flying mammals right out of the air.

As the children grew, so did the differences between their parents. Ron and Martha fought frequently—often over Joe. He was a prankish, sometimes exasperating boy, and Ron Slowinski was a perfectionist who expected strict order to be maintained in the household. He was a severe disciplinarian, more severe than Martha could tolerate. The couple divorced in 1974, when Joe was eleven. After the divorce, Ron moved into an apartment near the family house, so he could remain deeply involved in the children's upbringing. Three years later, he purchased a smart Cape Cod saltbox cottage near the Art Institute campus.

One of Joe's first major conflicts with his father was over religion. Even by the age of twelve, although he was too young to have more than a rudimentary grasp of science, Joe identified himself as a rationalist. One night at the dinner table he announced, "I don't believe in God. I don't believe in religion. I believe in evolution." Ron Slowinski was still devout at that time, and he insisted that Joe and Rachel go to church and Bible school. Joe eventually told his father, "I'll go to church because you make me, but I'm just going to sit there." Joe would remain attached to his father throughout his life, but on this point he always remained firm.

Martha now took the active role in supporting Joe's expanding exploration of the natural world. The first summer after the divorce, she took the children spelunking in Meramac State Park, a few miles south of Saint Louis. They rented a dark, mildewy WPA-era cabin in the woods and spent a week crawling through limestone caverns. Some of the caves had been developed for tourism, but Joe and Rachel wandered off the trails and found smaller caves hidden away in the woods, where they hunted for salamanders. Joe captured a blind, almost transparent grotto salamander. He had already discovered in himself an uncanny ability to spot cryptic species: Martha and Rachel had stared right at the creature and failed to see it.

When Joe was ready to enter high school, Ron and Martha took him out of the Kansas City public school system and sent him to Pembroke Country Day, a nearby prep school for boys. Joe won a full four-year scholarship there—based, in part, on an essay he wrote for the application entitled "Why Ecology Is Important." The following year, Rachel won a scholarship to Pembroke's sister school for girls.

The facilities at Pembroke Country Day were far more sophisticated than those at Missouri public schools: The science labs were equipped with the latest technology, the library was well stocked, the athletic facilities were luxurious. Yet, for Joe, the school's most fascinating resource was its menagerie. This pocket zoo had the usual assortment of small mammals and exotic birds; presiding over them all was a ten-foot-long boa constrictor—a lordly creature that Joe knew was at the top of the food chain in the menagerie's artificial mini-ecosystem.

As the school year was drawing to a close, Craig Maughan, Joe's biology teacher, asked the class for volunteers to take care of the animals over the summer vacation. Joe immediately laid claim to custodianship of the boa constrictor. Maughan said it was all right, if Joe's mother approved. Martha said that Joe could adopt the snake for the summer, provided he took full responsibility for it. He quickly set to work building a wooden cage in the basement, where the snake would stay.

By this point, Joe had already collected an impressive and diverse menagerie of his own. In his house, the animals always outnumbered their human keepers: there were dogs and cats (although, as domesticated animals, they were rather looked down on), birds, guinea pigs and hamsters, scorpions and spiders, and countless snakes, lizards, and toads. One of Joe's favorite pets was a tarantula named Mary: Everyone in the family became attached to it. Joe had found the spider crossing a highway in New Mexico, when he was visiting his father there, and brought it home in a Mason jar. When Joe approached Mary's cage, the spider would jump up and down, recognizing his scent. He used to let the fuzzy arachnid run all over his body. Mary fell off his shoulder one day and splattered to her death on the floor. Joe was depressed by the loss. He bought another tarantula at a pet shop, but he didn't bond with it as he had done with Mary.

On the last day of school, Joe brought the boa constrictor home. Soon a procession of neighborhood children was filing through the basement to view his spectacular acquisition for the summer. The cage Joe had built looked sturdy enough to persuade his mother to accept the exotic boarder, but the snake managed to break out within a week of its arrival.

One day, when Ron dropped Joe and Rachel off at the house after a visit, they found a note from Martha: "Welcome home, kids! I'm out shopping. Joe, your boa constrictor escaped." They ran downstairs to view the crime scene. Joe's cage was intact, but the bricks and stones he had used to secure the top lid were scattered on the floor. The muscular snake had knocked off the lid and bolted for its freedom.

A housewide search began at once, but the snake could not be found. Adapted to living in dark, warm places, it took up residence in the pipes and heating ducts of the house. The family was on a permanent state of alert, at the ready to capture the fugitive. Occasionally, they would see it slinking away, its tail disappearing into a vent, or its head would materialize for a shadowy instant under the drain in the kitchen sink. Sometimes, at night, ominous thumps would emanate from the core of the house; then days would go by without a sign of the serpent. When Rachel's guinea pig disappeared, she blamed it on the boa. One day, Joe and Rachel saw the snake crawling into a vent in the basement ceiling, with several feet of tail still hanging down. Joe cried to Rachel, "Come on, help me!" The two of them grabbed the end of the tail; the snake lifted them off the ground before it slipped out of their grasp and slithered back into its humid lair.

On a Saturday morning a few weeks later, Joe heard a suspicious noise and ran to open the cabinet under the kitchen sink. There was the boa, twined around the pipe. When Joe grabbed at the snake's head, it turned around and bit his hand. The boa got away, but one of its hundreds of tiny teeth broke off in Joe's finger, causing heavy bleeding and considerable pain. It was his first snakebite, and he was proud of it.

Martha was teaching a painting class at the Art Institute at the time, so Rachel called Ron. When they arrived at the emergency room, Joe announced triumphantly to the triage nurse, "I got bitten by a boa constrictor!" It took ten shots of Novocain and three surgeons to remove the needle-like tooth; the surgeon who succeeded in extracting it (by coincidence, a painting student of Martha's) performed the operation under a microscope.

The snake had the run of the house for months, until Joe finally caught it.

It's impossible to be certain on this point, as boa constrictors have had a long and flourishing career in children's cartoons, but Joe Slowinski's first encounter with the species was probably in the pages of *The Little Prince*, the classic fable by the French aviator Antoine de Saint-Exupéry, which was among the first books his parents gave him. Saint-Exupéry uses the snake as a symbol of fear, and to draw a parable of the literalness of the grown-up world. The adult narrator relates that when he was a child, he drew a picture of a boa constrictor:

> I showed the grown-ups my masterpiece, and I asked them if my drawing scared them. They answered, "Why be scared of a hat?" My drawing was not a picture of a hat. It was a picture of a boa constrictor digesting an elephant. Then I drew the inside of the boa constrictor, so the grown-ups could understand. They always need explanations.

The second drawing was reproduced in miniature on the last fifty-franc note issued in France, next to a portrait of Saint-Exupéry.

Obviously, no snake can eat an elephant, but boa constrictors and their near relations, the Asiatic pythons and South American anacondas, are ferocious hunters, capable of taking down and consuming animals much larger than themselves. First they disable the prey with a vicious bite that may sever an artery or tendon; then they wrap their thick, muscular bodies around the victim, crushing the life out of it and sometimes breaking bones, which makes it easier to swallow. They consume the prey whole, gaping their jaws, which are hinged with elastic ligaments that enable them to open their mouths to a width far greater than their bodies.

The largest species of the group, the reticulated python (*Python reticulatus*), is the longest snake on Earth; a specimen measuring more than ten meters, or thirty-three feet, was shot in 1912 on the island of Celebes in the Dutch East Indies (modern Indonesia). The heaviest snake on reliable record was a green anaconda (*Eunectes murinus*) killed in Brazil in 1960, which was just under twenty-eight feet long and forty-four inches in the "waist," putting its weight at well over five hundred pounds.

In 2005, a Dutch zoologist, Gabriella Fredriksson, published a paper about reticulated pythons in Borneo that hunt and eat Malaysian sun bears, which are ordinarily at the top of the food chain. After a three-year-old bear she was studying went missing, Fredriksson tracked down the radio collar it had been wearing—and found it in the grossly distended belly of a napping python. There is a real danger for the snake that such a supersized meal may rot in its gut before it can be digested and produce a lethal internal balloon of gas.

In 1929, a reticulated python ate a fourteen-year-old boy on the island of Salebabu, in modern Indonesia; in another well-attested incident, a youth running along a path in South Africa was cut down by a fifteen-foot-long rock python, which killed and partially consumed the boy while his neighbors tried to save him. There are myriad similar reports, but most of them melt into legendary mist when subjected to scrutiny. The people who live in the forest among these monstrous hunters often regard them as divinities; the animals' very hugeness endows them with an atavistic aura.

In his brilliant, encyclopedic study *Snakes: The Evolution of Mystery in Nature*, herpetologist Harry Greene conjectures that the instinctive human fear of reptiles may be rooted in our primordial ancestors' experiences witnessing their own kind being killed and eaten by enormous, terrifying creatures such as pythons and boa constrictors. These snakes, Greene writes, "teach us that ambush predation on large vertebrates arose prior to the origin of venomous species, and perhaps so did our mingled curiosity and fear of these creatures." The name *python* is derived from Pytho, a chthonic dragon of Greek myth, which was slain by Apollo in its lair in Delphi. The triumphant god established his oracle, the most famous in antiquity, on the site. In ancient vase paintings,

Pytho is always represented as a gigantic serpent. While no large constrictors have lived in Greece in the human era, it's possible that reports and even specimens of African pythons had reached the Mediterranean by classical times.

Boa constrictor is one of the oldest scientific animal names still in use, given to the species by Linnaeus himself, in 1758. Other names proposed for the huge serpent by rival taxonomists later in the eighteenth and early nineteenth centuries show the awe the species inspired in European naturalists, who were accustomed to snakes a fraction of the boa's size: *Rex serpentum* ("king of snakes"), *formosissimus* ("most beautiful"), *diviniloquus* ("speaking like a god," a reference to the Pytho legend). But Linnaeus's more down-to-earth, descriptive name (*boa* being the Latin word for an Italian water snake) stuck.

Boas, pythons, and anacondas are sometimes called "primitive" or "basal" snakes because they emerged early in the diversification of limbless reptiles—in other words, near the base of the hypothetical family tree, more than sixty-five million years ago. In the Oligocene epoch, around thirty million years ago, the "advanced" snakes—colubrids, vipers, and elapids—evolved a streamlined anatomy, having one each of the organs that come in pairs in most animals, including the basal snakes. Having one lung, one kidney, and so forth made the advanced snakes slimmer and faster; and with the evolution of venom, they were able to hunt even more efficiently than the bulky constrictors.

The question of the precise relationship between the New World basal snakes, the boas and anacondas, and the pythons of Africa and Asia has never been settled, and probably never will be. (Among other complications, there are three species of medium-size boas, typically not exceeding seven feet in length, on the island of Madagascar, off the east coast of Africa. Most likely, they are relicts of the geologic era when Africa and South America were joined in the conjectured supercontinent Gondwana.) Boas and pythons share many traits, primarily hunting by constriction; the principal difference between them is that pythons are oviparous, which means they reproduce by laying eggs, while viviparous boas give birth to live young.

Venezuelan herpetologist Jesús Rivas is developing a theory to explain this divergence. He believes that sixty to eighty million years ago, when modern South America was emerging as a distinct landmass, and before the upheaval of the Andes, there was a vast lake at the center of the new continent, where much of the fauna of the nascent Western Hemisphere evolved. According to Rivas's hypothesis, the boas, living in an aquatic environment, had no dry land where they could lay eggs, so natural selection favored viviparity. Boa constrictors produce large litters of young that are able to survive at birth.

The boa's strategy for survival may be of primeval origin, but it remains an impressively successful one; the species lives throughout the lowlands of mainland Central America. However, the snake was never known on Cozumel, an island ten miles off the east coast of the Yucatán peninsula, which has become one of the most popular resort destinations in Mexico. In 1971, a film crew from the mainland came to the island to shoot a costume drama about the Spanish conquistadors. To enhance the atmosphere of exotic legend, the filmmakers brought six boa constrictors to Cozumel for set decoration. At the end of the shoot, the wrangler in charge of the snakes let them loose in the wild.

Thirty years later, Cozumel was overwhelmed by their progeny. Set free in a closed environment thickly populated by the snake's preferred diet, yet void of the big cats and other large carnivores that prey on them on the mainland, the philoprogenitive boas numbered three thousand by 2003, when Jesús Rivas went there to make a film about this reptile irruption for National Geographic. He found an extraordinarily quiet jungle: The mammals that the boa preyed on, such as peccaries, raccoons, opossums, and coatis, were largely depleted; one of the three species of mouse that had inhabited the island for millennia had apparently become extinct. It was all too ironic: On the mainland, the boa constrictor is now an endangered species because of habitat destruction, while the prosperous resort of Cozumel, where, Rivas says, "ninety percent of the habitat is in good shape," is overrun by a plague of boas. The animal proved to be perfectly adapted to flourish in a micro-environment that had played no role in its evolutionary history.

Three years after he was there to make his film, Rivas said that Cozumel's boa constrictors had begun to shrink, not in population but in size: Now they rarely exceed four feet in length, whereas on the mainland they normally reach ten feet or more. The herpetologist hypothesized that natural selection might be favoring smaller individuals, which eat less. The females of these mini-boas aren't capable of bearing large litters; if the species continues to shrink in size, at a certain point it will become effectively infertile, reversing its population trend. The rapid transformation of Cozumel's ecology is a dramatic example of how fragile the balance is that underlies every ecosystem—even when it remains largely undeveloped.

One evening, at a meeting of the Heart of America Geology Club, a member brought in a fossil he had found on the banks of the Kaw River, just across the Kansas state line. Joe was exhilarated by the notion that he could go out and find fossils of his own: He was begging his mother to take him there even before they got home from the meeting. The next weekend they started exploring. Soon they found a favorite spot at a bend in the river near a little town called Bonner Springs, on Highway 7, a few miles downriver from the junction of the Kaw (the old, Indian name of the Kansas River) and the mighty Missouri, in the heart of Kansas City.

Martha would park the car by an old cast-iron bridge, and then mother and son would run down to the banks of the river, where the wide, flat sandbars still lodged fossil remains of the giant mammals that had roamed the Great Plains in the late Pleistocene era, ten thousand to twenty thousand years ago. Martha Crow later described the experience: "As we rushed down to the Kaw, hundreds of swallows, agitated by our arrival, would fly out from under the bridge to meet us, filling a beautiful blue sky."

The Kaw is a mellow, lazy stream that constantly exposes the fossilized antlers and bones that lie buried beneath it. Lewis and Clark camped on its banks in June 1804: At a court martial conducted there, two privates of the Corps of Discovery were convicted of stealing whiskey and getting drunk, and sentenced to floggings. The modern landscape is at once a muted reflection of the wilderness that Lewis and Clark found on their transcontinental trek and a distant yet distinct echo of the primordial natural world that thrived there before human migration. "There were so many fossils," said Crow, "and they were so easy to find and dig up, it would make anyone interested."

The participants in these Sunday outings to the Kaw varied; in addition to Martha, Joe, and Rachel, the group often included Rachel's friend Kim Moore and Sean Windsor, a classmate of Joe's from Pembroke Country Day. Windsor was also a scholarship student, which created a bond between the two boys at a school where most of the students came from rich families. One sunny afternoon on the Kaw, the two boys found a mastodon jaw, and Joe insisted on giving Windsor credit when it was exhibited at the Natural History Museum at the University of Kansas in nearby Lawrence. It was a rare moment of magnanimity: On another occasion, when Kim Moore found a large, handsome mastodon tooth, Joe became very annoyed with her for not giving it to him for his collection. "What use does she have for it?" he groused. (He got over it by his senior year, when he took Moore to the prom.)

Sean Windsor possessed the main requirement to be Joe's best friend: He shared his passion for exploring the countryside. Occasionally, the two boys would wander away from the river, seeking copperheads and nonvenomous snakes for Joe's ever-expanding menagerie. Joe was always lucky: It seemed that every rock he turned over had a snake under it. Completely fearless, Joe always tried to catch them with his hands. Every now and then, the boys did some forbidden botanical collecting: They found marijuana growing wild on the side of the road to Bonner Springs, which they picked and smoked before going to a Ted Nugent concert, but were disappointed when the herb failed to have the desired effect.

Joe's collecting habit was now deeply rooted, but the main motive guiding his outdoor exploration was a rapidly maturing scientific curiosity. He learned the scientific names of all the fossils he collected; Craig Maughan was impressed by his ability to "key out" the fossils, to determine what genus and species they were. By the end of the summer the boa constrictor had the run of his house, Joe had begun to develop a comprehensive profile of the biota (the flora and fauna) of the North American plains in the late Pleistocene era.

Two years after they joined the Heart of America Geology Club, Martha became an officer and took over the production of its mimeographed bulletin, and Joe became the science editor. She wrote up the minutes of the meetings, and he contributed short articles and illustrations. The son of artists, Joe had shown a remarkable talent for drawing since early childhood. His finely detailed renderings of fossils and animals, and imaginative re-creations of primordial landscapes with dinosaurs, enlivened the club's publication.

A few days after his fourteenth birthday, Joe wrote this short account of a fossil discovery for the publication:

A LUCKY FISHERMAN

By Joe Slowinski

> Like many fishermen, Francis Tully had his favorite spot—the lakes that punctuate the strip-coal-mining region, not many miles south of his home in Lockport, Ill., near Chicago. When the fish weren't biting, Tully passed his time by breaking open some of the rusty-brown ironstone concretions that littered the huge dumps of waste rock. Occasionally he would find a fossil fern inside one of the concretions, and one day in the early 1960s he found his first fossil shrimp, and, later, a worm. So far, however, his best find was a dragonfly entombed in a concretion. One day fishing was poor, so he cracked open a few concretions hoping to find a fossil fern or insect. But this time he had something unusual. What was the strange outline in the rock he held? The creature, if it was a creature, had an elongated, cigar-shaped body divided into segments. At one end was a spadelike tail and on the other a slender snout ending in a toothed

jaw. Behind the snout a bar crossed the body. The strange creature was about 5 inches long.

Tully was puzzled. Not knowing much about fossils, he brought it to the paleontologists at the Field Museum. They too were puzzled. This creature belonged to no known phylum. It was a stranger in the family of living things. However, even a stranger needs a name, so Dr. Eugene S. Richardson named it *Tullimonstrum gregarium*, or common Tully monster. Since then Tully has been back to the spot and now has around 4,000 fossils from the area.

This simple essay eventually proved to be as much about the storyteller as the story. Joe Slowinski would grow up to be both the lucky fisherman and the expert at the museum: the man who seeks out strangers in the family of terrestrial life, and the scientist who gives them their names.

Prairie Rattlesnake

Crotalus viridis viridis

The prairie rattlesnake, which ranges throughout western North America, is easily identified by the segmented tailpiece that gives it its name. The rattle is a loosely connected chain of knobs of keratin (an organic substance found in dead skin, horns, and fingernails), which are added one at a time, with every skin-shedding. Each knob is attached loosely to the younger one next to it; the segments at the end are the oldest. Specialized tail-shaking muscles rattle at a rate of fifty-five cycles per second, producing a hypnotic buzz that wards off predators. Rattlesnakes, like most New World venomous snakes, are pit vipers, so called because of the heat-sensitive depressions, or pits, between the eye and nostril on either side of the head, which enable the animal to aim its strike at warm-blooded prey. *Crotalus viridis viridis* is a common subspecies of the Western rattlesnake, *Crotalus viridis.*

Turning sixteen meant two things to Joe Slowinski: He was old enough to drive, so at least in theory he could go fossil-hunting on the Kaw River without making his mother take him; and he was now eligible to join a scientific expedition as a volunteer. Two years earlier, on a visit to the University of Kansas, Joe had met a paleontology professor who gave him a catalogue from the Earthwatch Institute, a nonprofit organization that recruits laypeople to become paying volunteers on field expeditions. The catalogue fired Joe's imagination: He brought it back to Kansas City and read it again and again. The Earthwatch programs were the intellectual equivalent of baseball camp with George Brett, the Kansas City Royals' All-Star third baseman, or a dance workshop with Mikhail Baryshnikov—a chance to play with the big boys, to do real science with real scientists. Joe was in the full flood of adolescence, eager for a chance to prove that he was ready to do a man's job; and the only adult activity he cared about was science.

There was a catch: The minimum age was sixteen. Impatiently he waited, and on the morning of his sixteenth birthday, he mailed off an application to Earthwatch. All the offerings in the latest catalogue intrigued him—a program to study endangered bird species in Hawaii, another to investigate the behavior of the spotted hyenas of Kenya—but Joe plumped for one close to home, a fossil dig in South Dakota called Mammoth Graveyard. A few months later, he got the news: He was one of eleven people chosen for the expedition and the only one to be offered a scholarship. Joe was elated. It would be his first long trip away from home on his own and his first organized scientific expedition. He was primed for a life-changing event.

His father drove him up to the dig, which was in the southwestern corner of South Dakota, about an hour south of Mount Rushmore. Like most divorced fathers, Ron Slowinski regretted the separation from his son and did what he could to stay close to him. So at Ron's instigation, they left Kansas City a few days early and made a father-son camping trip of it. Soon after they arrived in the Black Hills and set up camp, Ron drove into town to buy groceries, leaving Joe on his own. Ron felt a twinge of uncertainty about leaving his impulsive son behind and made Joe promise to be careful.

As soon as his father's car was out of sight, Joe headed into the woods—as always, on the lookout for wildlife. The rounded peaks of the Black Hills, darkly forested in Ponderosa pine, presented a wilder, more rugged prospect than the prairies around Kansas City he was accustomed to. Joe didn't expect to see a mountain lion or a black elk, but an opossum or a wild turkey, even an owl, would be exciting. When he was about a mile from camp, he turned over a large, flat rock and uncovered a healthy young prairie rattlesnake. The viper leaped up and bit him on the thumb.

Within minutes, Joe felt the debilitating poison take effect. The wound, bleeding slightly, throbbed with a hot, intense pain; soon his throat tingled with numbness; he felt a maddening thirst. Joe knew he had to get to a doctor as soon as possible—but how? He made his way back to the highway and tried to hitch a ride, but no one wanted to pick up a sweaty, disheveled teenager. In the end, he walked all the way to the rangers' station, a couple of miles away. By the time he arrived, his whole hand was swollen and darkening ominously. The pain was excruciating. The rangers rushed him in their patrol car to the hospital, where a doctor injected him with rattlesnake antivenom.

Joe had a severe reaction to the serum and went into anaphylactic shock, which required further emergency treatment. He would say afterward that, as far as he was concerned, the cure was worse than the disease. The antivenom Joe received in South Dakota was produced from the blood of horses that had been immunized by the periodic injection of small amounts of rattlesnake venom. Far from being a magic bullet, antivenom therapy commonly produces acute side-effects, the worst being a condition called serum sickness, which results in shivering chills, stabbing pains in the joints, and swelling and tenderness of the lymph glands.

The first antivenom for snakes was developed in Indochina by a French scientist, Albert Calmette, in 1895, as a cure for the bite of the Indian cobra. A few years later, scientists at the Instituto Butantan in

Brazil produced antivenoms for the New World venomous snakes, including the rattlesnake. To produce the antivenom, animals—usually horses and sheep—are injected with small amounts of venom sequentially, over a series of months. These small doses cause the animal to produce immune antibodies that neutralize the toxic effect of the venom. Then serum with the antibodies is extracted from the animals' blood and purified. It works exactly like a vaccine, with the animal serving as a surrogate.

Antivenoms fall into two main types: monovalent, which are effective against a particular species, and polyvalent, which counteract the toxic effects of bites from several species. Monovalent antivenoms are targeted very specifically, sometimes even aimed at subspecies. For example, different antivenoms have been developed for *Micrurus fulvius fulvius*, the harlequin coral snake, found throughout the eastern United States, and the Texas coral, *Micrurus fulvius tenere*, a very closely related subspecies. The most commonly used antivenom in North America is the polyvalent CroFab, a brand name for an antigen blend that is efficacious in treating bites by most species of the family Crotalinae, which includes rattlesnakes, cottonmouths, and copperheads. Because it is produced using live animals that are specially bred for the purpose, rather than in a chemical laboratory, antivenom is very expensive. One American pharmaceutical firm currently charges $979 per vial of polyvalent crotaline antivenom, and $1,428 for coral snake antivenom. A course of treatment against a large snake usually requires at least four vials, and sometimes as many as thirty.

In the developing world, prices are even higher and supplies are inadequate. It's sensitive stuff, which must be stored at a constant temperature not exceeding four degrees centigrade and administered by a doctor under controlled conditions that permit continuous monitoring of vital signs. For these reasons, even the most advanced, well-equipped expeditions rarely bring their own supplies of antivenom into the field; the standard procedure is to transport the bite victim to the hospital. And even under ideal conditions the treatment doesn't always work, as some antivenom therapies are simply less effective than others. For example, Ulrich Kuch, an elapid specialist at Johann Wolfgang Goethe University in Frankfurt, has reported that antivenom therapy often fails to save the lives of people who have been bitten by kraits.

The incidence of severely adverse reactions such as that which Joe experienced in South Dakota has been drastically reduced since then. Nevertheless, Joe made it a policy never to take antivenom again.

Soon he was fine and hard at work. Mammoth Graveyard is a sinkhole near the town of Hot Springs, which entombs the fossilized remains of more than a hundred Columbian mammoths, one of the largest species of the Elephantidae family that ever existed. Twenty-six thousand years ago, the gigantic, ungainly pachyderms slid one by one to a muddy death in a steep-sided pond, fifty feet deep and a hundred feet long. In 1974, a bulldozer digging the foundation of a new housing development struck a fossilized tusk. A geologist named Larry Agenbroad, who was working at a site nearby, was called in, and he immediately identified the specimen as a mammoth. The locality, now a National Natural Landmark, has proved to be the richest concentration of mammoth remains ever discovered in the Western Hemisphere. (By 2007, the tally at Mammoth Graveyard stood at fifty-five specimens, complemented by fossils of llamas, camels, and giant short-faced bears—all extinct in North America in the human era.)

In the summer of 1979, the team consisted of Agenbroad, four graduate students, and the volunteers: Joe, five other high-school students, and a seventy-one-year-old woman from Mississippi. They lived in tents by Meadowbrook Lake, near the site, and ate dinner around a campfire every night. During the day, Joe painstakingly excavated the square meter he had been assigned with a dental pick and a scraper. These "one-by-ones" were allocated randomly, but Joe drew the lucky lot. He found a complete mammoth skull, with both tusks intact: It was only the third complete skull of its kind that had been found on the site. It was a magnificent specimen—measuring ten feet from the rear of the cranium to the tip of the tusk—one that any scientist would have regarded as a major find. Joe had struck out George Brett, danced a pas de deux with Baryshnikov.

All rattlesnakes are native to the Western Hemisphere, and all are venomous. Snake venom varies enormously in toxicity among species and between individuals; the venom of the prairie rattlesnake that bit Joe is on the less potent end of the spectrum. It is stronger than that of the smallish sidewinder, *Crotalus cerastes*, yet nowhere near as lethal as that of the Mojave rattlesnake, *Crotalus scutulatus*, which has venom approximately as strong as an Indian cobra's.

Snake venom is radically modified saliva. The active ingredients are proteins, which constitute 90 percent or more of its dry weight (a breakthrough discovery in chemical herpetology made by Napoleon Bonaparte's brother Lucien, in 1843). Snake venoms are cocktails of hundreds, sometimes thousands, of different proteins and enzymes, which are constantly evolving. There are two main types of venom: neurotoxic, which disables the nervous system, and hemotoxic, which destroys soft tissue. To generalize, the venom of elapids, such as cobras, kraits, mambas, and coral snakes, is neurotoxic, while that of most pit vipers, including rattlesnakes, copperheads, cottonmouths, and the deadly Russell's viper of Southeast Asia, is hemotoxic.

In an article about snake venom for *California Wild*, published in 2000, Joe described the difference between the two types of toxin with a scientist's ghoulish humor: "Neurotoxic proteins are extremely effective at bringing about death with minimal disturbance to the body. Viperid venoms, on the other hand, are messy—tearing apart tissue and literally melting cells. So an elapid venom is more likely to kill you, but will leave a far prettier corpse."

Hemotoxins conquer the body in grisly cell-to-cell combat, killing muscle tissue as it inexorably advances through the victim's body; the medical term for this process, necrosis, is simply a transliteration of the Greek word for the state of being dead. Survivors of severe hemotoxic envenomation usually lose chunks of tissue or whole limbs. Costa Rica's first trained biologist, Clodomiro Picado, wrote a chilling description (quoted in Harry Greene's *Snakes*) of the effects of the hemotoxic venom of the fer-de-lance, which probably kills more people in the Americas than any other snake species:

> Moments after being bitten, the man feels a live fire germinating in the wound, as if red-hot tongs contorted his flesh; that which was most mortified enlarges to monstrosity, and lividness invades him. The unfortunate victim witnesses his body becoming a corpse piece by piece; a chill of death invades all his being; and soon bloody threads fall from his gums; and his eyes, without intending to, will also cry blood, until, beaten by suffering and anguish, he loses his sense of reality.

The neurotoxic venom of elapids, on the other hand, acts upon the nervous system, disabling muscle contractions and eventually causing total paralysis. Death from the bite of an elapid can come within minutes, caused by asphyxiation when the diaphragm, the muscle group that powers breathing, shuts down. The victim suffocates and dies as quickly as if buried alive.

In the late twentieth century, scientists began to realize that the venom of most snake species is actually a combination of the two types. Rattlesnakes are a good case in point. The Mojave rattlesnake was long known to be the deadliest species of the genus. In 1989, James Glenn and Richard Straight, two researchers at the Veterans Administration Hospital in Salt Lake City, studied the species' lethal venom and identified two discrete toxins, one producing "hemorrhagic and proteolytic activities" (causing massive bleeding and dissolving protein—in other words, a hemotoxin), and the other a classic neurotoxin, which quickly brought on fatal paralysis. Early in the course of their research, Glenn and Straight found that not all populations of Mojave rattlesnakes had the neurotoxin, which they called the "Mojave toxin," in their venom. The species has a wide range, from southern Nevada deep into Mexico, and from the Mojave Desert, in California, to West Texas. Mojave rattlesnakes in the deserts of the far West, in California and western Arizona, had the neurotoxin, but those farther east had only the expected hemotoxic venom. The scientists in Salt Lake City concluded that this gradual eastward expansion of the Mojave toxin was the result of interbreeding among contiguous populations.

Moreover, they found evidence that other species inhabiting territories that overlapped the range of the Mojave rattlesnake had begun to show significant amounts of the neurotoxin in their venom—including the prairie rattlesnake. In 1990, Glenn and Straight examined some prairie rattlesnakes from New Mexico and concluded that the snakes had acquired the Mojave toxin "as a result of some previous hybridization," or crossbreeding, with the Mojave rattlesnake.

In 1999, an eighteen-year-old reptile enthusiast in Hesperia, California, was bitten by a southern Pacific rattlesnake (*Crotalus viridis helleri*, which is very closely related to the prairie rattlesnake) he was trying to catch with his bare hands. Almost immediately, he experienced classic neurotoxic symptoms: difficulty in breathing, double vision, violent facial tics, and an inability to swallow or talk. It took thirty-five vials of antivenom to resuscitate him; two days later he was still unable to walk well. The victim also exhibited the common symptoms of hemotoxic poisoning: hemorrhaging and swelling. The doctors who treated him at the emergency room in nearby Loma Linda observed in a paper they published about the case that even if the snake had been misidentified and was really a Mojave rattlesnake, the incident was still significant because envenomation demonstrating both neurotoxic and hemotoxic effects "has not been reported previously from southern California."

Over the past twenty years, scientists have found evidence of neurotoxins in the venom of ten of the fifteen species of rattlesnake native to the United States. In many cases they have been trace amounts, but recently even *Crotalus horridus*, the timber rattlesnake of the eastern and central United States, which inhabits a territory far from that of the Mojave rattlesnake, has proved to have lethal amounts of the Mojave toxin.

In 1998, during a service at the Rock House Holiness Church in northeastern Alabama, John Wayne "Punkin" Brown Jr. was preaching with his own three-foot-long timber rattler when the viper sank a fang in his finger. Snake handling is a controversial ritual practiced by a few Pentecostal Christian congregations that is based upon Jesus' prophecy in the Gospel of Mark that his followers "shall take up serpents; and if they drink any deadly thing, it shall not hurt them." According to witnesses, Brown "emerged from behind the pulpit, stepped down onto the church floor, and toppled over." Brown was dead in ten minutes—long before the hemotoxin normally associated with timber rattlesnakes could have taken lethal effect. (Brown's quick death might have been the result of anaphylactic shock—an extreme allergic reaction caused by his previous exposure to the snake's toxin.)

Scientists offer competing explanations for this epidemic rise in neurotoxic venom in rattlesnakes. Many believe it's a proof of rapid molecular evolution, as suggested by Glenn and Straight—the result of localized crossbreeding that began with Mojave populations in the American Southwest and spread eastward. Other scientists doubt that such a significant evolutionary change could have occurred within the space of a few decades. One imaginative explanation of the phenomenon is the theory that venomous snakes and their prey are locked in an evolutionary "arms race": The prey are constantly developing resistance to the venoms of the snakes that hunt them, and in response the snakes evolve new toxins to kill the prey.

This scenario doesn't require the emergence of new species or subspecies. Rather, it is something of a micro-retrofit on existing species, in which natural selection works with lightning speed to make small biochemical adjustments that maximize the odds of successful hunting—and, therefore, of survival. The genes of small mammals and other prey that survive the bites of predators are passed on to their offspring; the same process is at work for snakes, which develop new, deadly toxins that overpower the biochemical defenses of their prey: hence the metaphor of an arms race. The concept is also called the Red Queen Hypothesis, a reference to the scene in *Through the Looking-Glass* in which Alice runs furiously to keep up with the Red Queen, only to find at the end that she's in the same place where she started.

Natural selection might also be exerting an influence over rattlesnake behavior. In 2001, the *San Francisco Chronicle* interviewed Joe Slowinski for a story with the alarming headline RATTLESNAKE DANGER GROWS AS MORE SERPENTS STRIKE WITHOUT WARNING. The article reported that rattlesnakes were rattling less than they used to—a phenomenon attested by biologists, park rangers, and hikers. Joe told the *Chronicle*, "We have lots of old accounts from naturalists who found snakes by listening to the rattle, but you don't see that now. We found some rattlesnakes last weekend fifty yards from a fire trail near Mount Tam. We came right up on them and never heard a rattle."

The prevalent theory to explain the rattle adaptation is that it evolved as a defense mechanism, to ward off big mammals such as coyotes, which prey on snakes. Joe thought that natural selection now favored silent rattlesnakes, because rattling behavior attracted the attention of the snake's main predator in the modern West: the complex, unpredictable mammal known as *Homo sapiens*, which is making ever deeper incursions into rattlesnake territory. Joe explained, "Snakes that never rattle are more likely to survive human predators. For decades in the American West, people have killed rattlesnakes. But if we don't hear them, then they usually go unnoticed."

The vast majority of snakebite victims in the United States are very likely to survive, provided they follow the cautionary advice of park rangers, wilderness guides, and wildlife Web sites, and seek immediate medical care. Estimates for the number of venomous snakebites in the United States range from one thousand to four thousand yearly, yet the number of deaths is fewer than ten. Nonetheless, there's no doubt that rattlesnakes pose a greater danger to people now than they did in 1977, when Joe turned over that rock in the Black Hills.

Joe came back from South Dakota full of enthusiasm. He loved the camaraderie of camp life and the gritty physical contact with the earth; he had added impressively to his mastery of field technique; and he was justifiably proud of having made a significant scientific discovery. The mammoth skull he found provided a window onto the late Pleistocene, when giant mammals roamed the Great Plains—a primordial vision of Joe's homeland thousands of years before the first humans arrived.

Ever since he had joined the Heart of America Geology Club, Joe had nurtured a vivid fantasy of himself as a scientist in adulthood; the experience of excavating the Columbian mammoth skull transformed it into a commitment. It had been a major test of skill to extract the massive fossil from the earth—and, more important, an intellectual challenge to interpret it correctly. In the fall, in an interview with his high-school newspaper, *The Hilltop*, Joe revealed how the experience had shaped his dreams of the future: "I am an outdoorsman. Paleontology gives me a chance to be outdoors and have experience with animals. I see myself traveling to various countries in search of fossils, animals, and cultural aspects of life."

Yet when that prairie rattlesnake, roused from its cool slumber under a rock, flew at Joe and sank its fangs into his flesh, releasing a debilitating dose of venom, he experienced the most exhilarating, visceral challenge of his life so far. None of the animals he had interacted with up to that point had posed any serious danger to him, not even the fugitive boa constrictor. In South Dakota he had met a living adversary poised for mortal combat—a pure, astonishing expression of the power of nature. Joe's engagement with venomous snakes would soon become as much a passion as a cerebral pursuit.

American Copperhead

Agkistrodon contortrix

The American copperhead is the most common venomous snake in North America, but its relatively weak venom virtually never results in human death. Its scientific name is a fanciful description of the serpent's backward-curving fangs: The genus name is derived from the ancient Greek words *agkistron*, for fishhook, and *odon*, for tooth; *contortrix* means twisted or turned back. Copperheads are diurnal in cool spring and fall weather, but nocturnal in the heat of summer; they hibernate in winter. They are a social species, frequently found sunning in groups near hunting sites and watering holes. They overwinter in communal dens with other species of snakes, including timber rattlesnakes and black rat snakes.

Joe's formal education as a scientist began in 1980, when he enrolled at the University of Kansas, just across the state line. After his summer in South Dakota, everyone—including Joe—assumed that he would major in paleontology. As soon as he arrived in Lawrence, he got a work-study position supervised by a paleontology professor named Larry Martin, a frequent guest speaker at the Heart of America Geology Club. It was a good job: Joe was responsible for identifying and interpreting fossils in the backlog of the university's Natural History Museum.

The collection was housed in Dyche Hall, an eccentric, turn-of-the-century Romanesque pile built of local limestone. Working in the building's basement, a dim warren of offices, labs, and musty storage bins, Joe met another freshman named Stanley Rasmussen, the son of a paleontologist who had once taught at the university. Tall and rangy, with a long, solemn face that belied a sly sense of humor, Rasmussen was as smart as Joe and shared his passion for the outdoors; he had also picked up a lot about paleontology from his father. The two young men soon became fast friends.

The first weekend he could borrow a car, Joe took Rasmussen to Bonner Springs to look for fossils. The Kaw was at the lowest level Joe had ever seen; sandbars that had always been submerged were now exposed. He and his new friend made a spectacular haul that day: three bison skulls, three deer skulls, a mastodon pelvis and part of a mastodon jaw, moose antlers, a coyote ulna, a bit of fossil turtle shell, and a fish vertebra—a compact Pleistocene zoo in stone, well-preserved in the soft river silt. It was exhausting, back-breaking work: dragging hundreds of pounds of fossils up the riverbank to the road, loading them into their friend's car, and scraping themselves up in the process. They dropped one of the bison skulls, breaking it in half.

Joe sent a letter home detailing what they had found and explaining how the expedition had ended, a story that must have exacerbated any maternal misgivings Martha Crow may have had about sending her excitable seventeen-year-old son into the world on his own: "As we were collecting on a sandbar near the railroad bridge, a man with a rifle (on the opposite side) began shooting in our direction. Apparently, he didn't see us. Each time he shot we could hear the bullet whistle past us. Then we wisely decided to leave." When they returned to Lawrence, Joe and Rasmussen donated most of the fossils to the museum—but kept the best specimens for themselves.

Joe soon came to feel at home in Lawrence, a college town that was at once more easygoing and more intellectually stimulating than Kansas City, where business ruled. He was shy with girls, preferring male camaraderie; he quickly developed a penchant for drinking beer and playing pool at the bars that catered to KU students. Most weekends, he and Stan Rasmussen would collect fossils and chase snakes. They were both Clint Eastwood fans and could recite the dialogue of the Dirty Harry movies by heart. Above all, they loved Warner Bros. cartoons; as they bounced through the woods, they would talk to each other in character as Bugs Bunny, Elmer Fudd, and Yosemite Sam. Their all-time favorite cartoon was "Lonesome Lenny," a Tex Avery riff on John Steinbeck's *Of Mice and Men*, about a feeble-minded dog at a pet shop that squeezes his best friend, a squirrel, to death. The boys named one of their most beloved snakes after the cartoon's title character.

Joe's collecting habits put some strain on his living situation. Before classes started, he had been too lazy to join the matchmaking system for assigning roommates, so he was arbitrarily paired with a Japanese graduate student. In spite of Joe's childhood experience of living in Kyoto, the two didn't get along. Joe soon filled their room with fossils and wildlife; he had a vivarium that was always crawling with snakes, lizards, and spiders that he and Rasmussen had collected on the Kaw. In a letter home, Joe wrote: "Yukihiro is irritated by the build-up of bones in our room."

That spring, Joe and Rasmussen made several major finds, some of which are still on exhibit at Dyche Hall, such as a complete, well-preserved fossil mastodon skull Joe dug up from a sandbar on the river. One of his earliest scholarly papers, published in the *Transactions of the Kansas Academy of Science*, described the ulna of a giant beaver that Stan Rasmussen found on the Kaw. Joe also made a significant archaeological discovery his freshman year: On a solitary hike he found the remains of a bison kill, a place where early Plains Indians had rounded up a herd of bison and slaughtered them. An archaeologist at the university expressed interest in excavating the site.

Joe was getting a reputation: His classmates at the University of Kansas called him the King of the Kaw. He and Rasmussen dreamed of taking their paleontological pastime to a professional level: Rasmussen knew about a place in Montana where there was a complete fossil Triceratops, one of the largest dinosaurs of the late Cretaceous period, just waiting to be dug up. They formed a bold scheme to excavate the great reptile's skeleton during their summer vacation and sell it to a museum; a complete Triceratops specimen was worth hundreds of thousands of dollars. Yet it was doubtful whether a pair of teenage students would have been able to make a convincing legal claim to ownership, even if they had succeeded in extracting such a huge, complex fossil by themselves. In the end, they abandoned the project after Rasmussen got a summer job at a Boy Scout camp in New Mexico. Anyway, by this time Joe's focus was turning from fossils to snakes.

Although he chose to attend the University of Kansas mainly because of its proximity to home, and to some degree because of his friendship with Larry Martin, it was an excellent place to study herpetology. Among the faculty stars were William Duellman and Linda Trueb, who between them (and together) had written some of the basic texts in the field, and emeritus professor Henry Fitch, an internationally recognized expert on copperheads who would play an important mentoring role in Joe's education.

Improbably, the state of Kansas has fostered a noble tradition of herpetology. Edward Drinker Cope, America's greatest herpetologist, did his first fieldwork in Kansas in 1871, under the auspices of an official U.S. government survey. Just two years before Joe arrived at the University of Kansas as a freshman, another giant of the field, Edward Harrison Taylor, had died in Lawrence at the age of eighty-nine.

A native of De Kalb County, Missouri, Edward Harrison Taylor studied paleontology and zoology at KU. After graduation, he went to the Philippines, then an American dependency, where he became an administrator at the Bureau of Science. Taylor's field experience as a herpetologist ranged across the Malay Archipelago, Thailand, Mexico, and Costa Rica. (He also served as a spy for the American government in Siberia in the early 1920s, and in Java and India at the end of World War II.) Taylor described more than five hundred new species of amphibians and reptiles, most of them specimens he had collected himself. Eventually he returned to the University of Kansas, where he served for a time as the curator of herpetology at the Natural History Museum.

That fall, Joe and Rasmussen caught Penny, a gravid (pregnant) copperhead, and brought her back to the dorm, where she delivered a squirming litter of seven baby snakes. Most snake species simply abandon their newborn young or, if they are oviparous, leave eggs behind immediately after they've laid them. Yet some pit vipers exhibit a greater degree of maternal behavior than most snakes. After they give birth, copperheads watch over their brood for a period of a week or more, until the young are able to strike out on their own and fend for themselves.

Juvenile copperheads have a highly specialized hunting strategy: They are born with a bright yellow tail tip, which they use to attract prey. The tiny snake coils itself on the forest floor and extends its tail tip in the air and wiggles it, to mimic the movements of a worm or grub. When a small frog or lizard approaches to investigate, the copperhead strikes, envenoming and subduing the prey, and then eats it. It's only a transitional tactic; as soon as the snake is big enough, it shifts to hunting rodents. More is known about the copperhead than almost any other North American snake, thanks in large measure to a comprehensive study of the species by Henry Fitch, published in 1960, which Harry Greene praised as "without peer in the literature on snake ecology."

Joe and Rasmussen eventually released the baby copperheads in the ravine where they had found Penny, but they decided to keep her through the winter as a pet. When she came down with a case of mouth rot, they nursed her back to health: Rasmussen held her down on the table while Joe swabbed out her mouth with hydrogen peroxide and squirted it with liquid tetracycline. In the spring, they returned Penny, healthy again, to her old haunts.

It was inevitable, living among venomous snakes as he did, that Joe would eventually get bitten by one of his pets. His junior year at the University of Kansas, as he was dropping a copperhead into a pillowcase, the snake caught a fang in his thumb. He lanced the wound with a razor blade and had a friend take him to the hospital. The thumb turned purple and swelled to a comic hugeness, until, Joe said, it looked like a Fred Flintstone thumb injury. "It felt like someone took a mallet and pounded my thumb with it, and kept pounding," he told Rasmussen the next day, when his friend came to pick him up at the hospital. They drove straight to the woods for more snake hunting. Joe could only use one hand, because his thumb was still swollen and swathed in bandages, but he still managed to catch another copperhead that day, left-handed.

What compels someone, even an impetuous twenty-year-old college student bursting with curiosity, to go running back to repeat an experience that has made him seriously ill? Ordinary people develop a fearful aversion to snakes after such an encounter, but there was nothing ordinary about Joe's attraction to reptiles.

The fear of snakes is an ancient human instinct. Most people, even if they have had no unpleasant experiences with the creatures, tend to fear them viscerally: A garter snake swiftly slithering through tall grass can give a grown man a scare. According to medical researchers at Purdue University, ophidiophobia, a neurotic fear of snakes, is the most common specific phobia in the United States. Even great scientists succumb to irrational abhorrence: Linnaeus himself described reptiles as "foul, loathsome beings." Roy Chapman Andrews, one of the boldest scientific explorers of the twentieth century, who led historic expeditions throughout Asia for the American Museum of Natural History, confessed, "I dislike reptiles intensely. I don't know why, but I just do. My dislike isn't fear. It is an instinctive loathing."

Harry Greene's hypothesis that the dread of snakes may be hard-wired in the brain, an atavistic instinct rooted in primordial encounters between primates and deadly snakes, finds support in the ancient cultural record. In the Book of Genesis, after the serpent, "more subtle than any other wild creature that the Lord God had made," seduced Eve to taste of the Tree of Knowledge, God in his wrath put a curse on the creature: "Upon your belly you shall go, and dust you shall eat all the days of your life." God declared eternal hostility between humankind and snakes. Addressing the serpent, he said, "I will put enmity between you and the woman, and between your seed and her seed; he shall bruise your head, and you shall bruise his heel."

A heavy sentence: whether the author of Genesis reflected a prehistoric fear of snakes, based upon millennia of observations of their shocking predatory behavior, or helped to promote the fear by placing the curse of God upon them, in the Judeo-Christian tradition the serpent is the embodiment of evil. Its staring eyes with slitted pupils and darting, forked tongue became common features in artistic representations of demons—foul spirits outside the dominion of God. The legend of Saint Patrick driving the snakes out of Ireland (which in fact never had any snakes to begin with) is a metaphor describing the triumph of Christianity over the ancient pagan religions.

Yet in cultures beyond the influence of the Hebrew Bible, snakes are often venerated. In the ancient Greek cult of Asclepius, snakes were used in healing rituals, a tradition that survives in the caduceus, a winged rod entwined by a pair of serpents (originally, the staff of the Greek god Hermes), which has been the symbol of medicine for centuries. In Aztec Mexico, Quetzalcoatl, the feathered serpent, was one of the most beloved gods of the pantheon, the bringer of knowledge and skill—subtlety in the positive sense.

Like every outcast, the snake has always attracted passionate admirers, particularly among outsiders, those who flout the social order. In the American Revolution, the native rattlesnake was a defiant symbol of resistance. One of the most popular American flags before the adoption of Old Glory was the Gadsden flag, with the image of a coiled timber rattlesnake and the legend DON'T TREAD ON ME. Ben Franklin (an outlaw at least in the eyes of the British) opposed the bald eagle as the new nation's symbol, calling it "a bird of bad moral character" and proposed in its place the rattlesnake: In a letter to the *Pennsylvania Journal*, published in 1775, Franklin wrote, "She never begins an attack, nor, when once engaged, ever surrenders: She is therefore an emblem of magnanimity and true courage."

In the country's westward expansion, a new myth, that of the fearless snake wrangler Pecos Bill, was created, which melded the supernaturally strong heroes of European legends with the courageous Indian snake-handlers whom the pioneers had encountered—like the priests of the Hopi snake cult Joe witnessed when he was a boy. Pecos Bill was the epitome of all manly virtues in the American West, the strongest, bravest cowboy of them all. When he had a run-in with a fifty-foot rattlesnake, Bill let the snake bite him three times before he thrashed it, just to make it a fair fight. After that, the rattlesnake was his living lasso; instead of a horse, Pecos Bill rode a mountain lion. Thus he tamed the two most dangerous carnivores of the West.

The myth of man triumphing over snake received a feminine twist in the weirdly glamorous person of Grace Olive Wiley. Wiley was a curator at the Brookfield Zoo, one of the nation's largest, in suburban Chicago. Originally a librarian in Minneapolis, she got the job at Brookfield in 1933 after she offered to give the zoo her personal collection of 350 reptiles if they hired her. Wiley believed that venomous reptiles could be tamed; she handled them casually with her bare hands every day. In an article for *Natural History* magazine, published in 1937, she claimed that cobras, mambas, fer-de-lances, coral snakes, moccasins, copperheads, and rattlesnakes (among others) could be handled safely and liked being stroked. "Somehow they know very, very soon that I am friendly and like them," she wrote. "They appear to listen intently when I stand quietly at their open door and talk to them in a low, soothing voice. In some unknown manner my idea of sympathy is conveyed to them."

She was a disastrous curator: The animals in her charge were always escaping. In all, she let nineteen snakes get loose, most of them venomous species. When the zoo's director, Robert Bean, finally fired her, *Time* did a story. Defiant, Wiley told the reporter, "I hate to say it, and I know some persons who don't like snakes are very nice persons, but Mr. Bean was frightened, and frightened persons will exaggerate. I do not feel I was guilty of carelessness. I just forgot, simply forgot, to close the door to the cobra's cage after I cleaned it." Later, according to a biographical sketch by James Murphy and David Jacques published in the *Herpetological Review*, Wiley set up a roadside attraction in Cypress, California: a petting zoo where children could play with king cobras, kraits, copperheads, and rattlesnakes. She died after a cobra bit her while she was wrangling the snake for a photographer from *True*, a popular men's magazine. She had some antivenom on hand, but before she could take it, someone accidentally broke the vial.

The myth of human conquest of the snake kingdom survives uncertainly in the rattlesnake roundups of today. These orgies of ophidiophobia occur in New Mexico, Oklahoma, Kansas, Alabama, Georgia, and Pennsylvania, but most of the old-style roundups are held in Texas. They originated with annual sweeps by ranchers who wanted to reduce the population of rattlesnakes, which they believed posed a danger to their cattle and horses; today, they're more like gothic country fairs. The festivities include snake-sacking contests and daredevil acts: A man is zipped up in a sleeping bag full of rattlers and rolls around, another performer coils an angry, rattling diamondback on top of his bald head. The acknowledged champion today is Jackie Bibby, who holds four world records for his stunts. A native of Rising Star, Texas, Bibby claims to have been bitten by rattlesnakes eight times since he started performing with them in 1969. He told a newspaper reporter, "I'm an egomaniac with an inferiority complex, and I'll do anything to get attention."

The largest event of this kind is the Sweetwater Rattlesnake Roundup, which has been held every March since 1958, sponsored by the Jaycees to support local charities. The centerpiece of the event is the snake pit, a circular cavity filled with hundreds of live rattlesnakes that are brought in by snake hunters who sell them for about four dollars a pound. Snakes are continually hauled out of the pit, slaughtered and skinned, then battered and fried: A portion of rattlesnake and french fries sells for five dollars.

Another popular Texas rattlesnake roundup is held in Freer. It began in 1965 as the Freer Oil-O-Rama, where people came to look at oil-drilling equipment, drink, dance, and watch the Miss Oil-O-Rama beauty contest. Many visitors took a greater interest in the rattlesnake show at the Texaco station down the road from the Oil-O-Rama, so the event gradually transformed into a roundup based on the successful Sweetwater formula.

Snake roundups these days are little more than reptilian holocausts. Professional hunters roust rattlesnakes by spraying gasoline into their dens, killing many other small animals and making the site uninhabitable for years. The roundups continue an old tradition of persecuting rattlesnakes, until recently abetted actively by park rangers, which has driven the timber rattlesnake—the animal Ben Franklin proposed for America's national symbol—to the verge of extinction. Animal-rights activists now show up routinely at the roundups to handcuff themselves to fences and shout slogans—a quixotic tactic against events that bring hundreds of thousands of dollars into the communities that stage them. Snakemen like Jackie Bibby walk right past the protesters without giving them a glance.

Of course, Joe Slowinski lived in a different moral and intellectual universe: He was devoted to the preservation of reptiles. In 2000, he wrote in *California Wild*, "The large-scale killing of venomous snakes under the pretense of saving human lives is indefensible, and many of the bites that occur each year befall the misguided people participating in rattlesnake roundups, annual events that occur across the United States to rid natural areas of venomous snakes."

Yet as dissimilar as Joe was to outlandish Grace Wiley and a carnival artiste like Jackie Bibby, it wouldn't be quite true to say that he had nothing in common with them. By the time he was studying biology at the University of Kansas, Joe had discovered in himself a consuming, deep-lying fascination with venomous snakes. George Zug, a senior curator at the Smithsonian who would later be the codirector of the Myanmar Herpetological Survey, Joe's research project in Burma, believes that this primal connection with snakes was an integral part of Joe's scientific career. Zug said, "Joe was a snake freak. His extreme enthusiasm for snakes drove the carelessness in his handling of them—but it also drove his appetite to study them."

Joe himself was baffled by his enthusiasm. When he was interviewed on the Discovery channel in 2001, he said, "I can't even explain why I'm interested in venomous snakes. I've been that way ever since I was a little kid. I always loved snakes, but for some reason the sight of a rattlesnake or a copperhead really got me excited. There's no way I can explain that. It can't be a gene: How could a gene like that ever survive natural selection?"

While he was a student at KU, Joe's burgeoning passion sent him ever farther afield in his search for herps (biologists' argot for reptiles and amphibians collectively; those who pursue them are known as herpers). Now with a car of his own, a homely Volkswagen Beetle, he drove to Taos after his junior year to visit his father at his usual summer retreat, accompanied by a large *Crotalus atrox*, the western diamondback rattlesnake. Then Stan Rasmussen joined him, and they headed farther west, into the Sonoran Desert, where both young men felt overwhelmed by the mystical power and beauty of the landscape. After hiking Organ Pipe National Monument, they dipped into Mexico, where they collected rattlesnakes and a fine Gila monster, the legendary venomous lizard of the Southwestern desert. It would have been illegal to collect the animal in the United States, and it was probably illegal to bring it back from Mexico. When it came to reptiles, Joe never let rules and regulations stand between him and a good specimen.

Joe's dedication to field research had become intense, but perhaps more significant was his transformation into a disciplined thinker who excelled at the demands of the academic life. His classroom style as an undergraduate wasn't flashy, but his intuitive comprehension of complex issues of evolution and ecology was profound. When he got a 95 on the midterm exam of a tough biology course, the professor wrote at the top: "This is a very good exam. I had no idea what to expect from you since you seem reluctant to participate in discussion. Also, you frequently appear to be bored with the lectures. Have you had much of this material previously?" No, Joe hadn't "had" the material, in the sense of having studied it in a course; he had read and thought about the issues on his own and talked about them with Stan Rasmussen while they walked down rivers and through woods.

By his senior year, Joe had made a firm commitment to go to graduate school in biology. His weekend wanders with Stan Rasmussen were now informal scientific field expeditions, and they knew what they were doing. In the spring, they discovered a den of timber rattlesnakes sharing a rocky ledge with a troop of sociable copperheads in an exurban area slated for a future housing development. It was far from the city, in rural Johnson County, Kansas, but it already had a Kansas City street address. Joe called on Henry Fitch and brought him to see what he had found.

Fitch was amazed by the density of the snake population. He taught Joe how to tag them by clipping the subcaudal scales, those on the underside of the tail. Each specimen is marked in a different pattern, like Morse code, so that it can be identified later if recaptured. The two of them, the retired professor and the aspiring scientist, caught thirteen rattlesnakes and eleven copperheads that day; all were weighed, measured, marked, and released. "I was very pleased to find a youngster so seriously interested in studying snakes," Fitch recalled many years later. "I was anxious to show him what I knew about capturing and handling them."

(Three years later, when Joe was living in Florida, he came across a news story in the *Kansas City Star* reporting that construction was going ahead at the site. He again enlisted the aid of Henry Fitch and Stan Rasmussen, the latter of whom by then had returned to Lawrence with a law degree from the University of Colorado, to try to save the snake den. They got nowhere, because the timber rattlesnake was not a threatened species in Kansas.

In 1984, his senior year at Kansas, Joe applied to and was accepted by the Ph.D. program in biology at the University of Miami in Coral Gables, Florida. He could have stayed in Lawrence for his graduate study, but he had heard great things about the chairman of the biology department at Miami, a herpetologist named Jay Savage, and the Organization for Tropical Studies that Savage had cofounded in Costa Rica in 1962. Joe already knew that he wanted to work in the tropics; that was where most of the venomous snakes were.

The winter of 1984 was one of the coldest in memory in Kansas. One weekend, Joe and Stan Rasmussen had planned an excursion to the Squaw Creek National Wildlife Refuge in Mound City, Missouri, to see the bald eagles the place was famous for, but when they phoned ahead, the rangers told them there were no eagles, on account of the cold. Wintering eagles, Joe knew, required open, unfrozen water, and there was none at Squaw Creek, with its shallow, manmade swamps created to lure the birds in warm weather. So he and Rasmussen went to the Kaw, in the hope that there would be some open water there. They were in luck. He wrote his mother an account of what happened:

> The morning we went out it was very cold and snowy. The Kaw was frozen solid, so we walked on top of it. We walked for several miles without success, and were ready to give up, when I decided to scan a large grove of cottonwood on the north bank with my binoculars. There in the branch of a cottonwood I saw a black form with a bright white head. Twenty feet from it was another eagle. It was much more exciting than seeing them at Squaw Creek, because here they were truly in the wild. We tried to approach them, but they flew away before we could even see them with our unaided eyes. We walked over to where they had been, and sure enough, there was a thin strip of open water amidst all the ice.

A few months later, on their graduation day, Joe and Rasmussen went out for a celebratory snaking expedition with Martha and Henry Fitch. It seemed to the others that every rock Joe turned over had a snake underneath it; he bagged more than the rest of them combined. It was uncanny.

Monocled Cobra
Naja kaouthia

The monocled cobra takes its name from the distinctive mark on its hood, a blobby O-shaped pattern that nineteenth-century herpetologists thought resembled a monocle. The hood is a defensive display, which the snake flares when it is being menaced by predators. It is formed when elongated anterior ribs, near the head, extend laterally, stretching the skin. The cobra always hoods before it strikes. Its fangs are fully hollow, permitting a quick injection of the snake's highly toxic venom, which rapidly paralyzes its prey, usually birds and small rodents. *Naja kaouthia* ranges from northeastern India and southern China throughout mainland Southeast Asia. It is not an endangered species, but is under stress from excessive hunting by humans—for food, the leather trade, and the traditional Chinese medicine market, which prizes it as an aphrodisiac.

In Jay Savage, Joe found both an intellectual mentor and a role model. The genial, brilliant biologist would become his mentor in an official sense as his Ph.D. supervisor; he also served as a scientific father figure throughout Joe's life. The two resembled each other physically: good-looking men who perpetually wore a wide smile that revealed a gap between the upper incisors, giving their faces a sunny, boyish aspect. Savage embodied the classic herper ethos: a confident machismo mellowed by a love of nature, the outdoor life, and the camaraderie that goes with it, governed by a creative mind rigorously trained to rationalize the complexity of life on Earth.

Savage would later say, "Joe was obviously a very bright guy, and he soon revealed himself to be a serious thinker, with great mathematical abilities and a talent for systematics and evolutionary biology." Systematics is the study and classification of organisms with the goal of establishing their evolutionary histories and relationships—an abstruse field that became Joe's specialty at graduate school. Yet Savage also recognized that his new student, a young twenty-one, had an impulsive streak that might get him into trouble. Joe was accustomed to playing the part of the precocious youth, petted by his elders; if he was going to survive his first year at graduate school, he needed to grow up fast and develop some personal discipline.

It was Savage's unofficial policy to assign advanced students to newcomers to play the part of big brother, acting as an informal surrogate for Savage himself. He matched Joe with Brian Crother, a tanned, rakishly handsome Californian. Crother described Joe when he arrived in Coral Gables as "a skinny, white-bread Midwesterner, fresh from K.C., with the beady blue eyes of a snaker." Crother was seven years older than Joe, having taken a few years off before he began his graduate studies, but Savage's intuition that the two men would get along well was quickly proved right: In no time, they were going on weekend trips to the Everglades together to collect herps, just as Joe and Stan Rasmussen had done in the woods of Kansas.

One of their first excursions was to an auction at the Miami Serpentarium, a tourist attraction owned by a man named Bill Haast, who billed himself as the most snakebitten man in the world. The Miami

Serpentarium was a relic of the golden era of roadside attractions in Florida: A thirty-five-foot-tall stucco sculpture of a coiled, hooded cobra welcomed visitors to the park. (After he closed the business, Haast donated the sculpture to South Miami High, whose sports teams were known as the Cobras, but the snake's head fell off on the highway when it was being transported to the school.) Beneath his flamboyant façade, Haast had serious intentions: The Serpentarium was for many years the largest producer of snake venom for scientific purposes in America. After thirty-eight years in business, however, Haast had decided to shut the place down and sell off his collection. Pythons were going cheap, so Joe bought a ball python, whose name comes from its defensive adaptation of rolling itself into a tight sphere with its head tucked into the core.

When Joe first arrived in Miami, many people wondered why he had chosen to study there, rather than to continue at KU. Before long, however, the lab had nicknamed him Duellman's Revenge, in recognition of the friendly rivalry between Jay Savage and his colleague in Kansas. The name fit him: The wild exploits of the new student from Kansas City were soon being gossiped about throughout the university and beyond. As Brian Crother put it, "A legend was in the making."

One infamous escapade began when Brian Crother and another graduate student took Joe to visit Zoological Imports, a company that specialized in exotic animals, selling everything from tarantulas to ocelots. Most alluring of all, the shop had a "hot room," which was devoted to venomous animals. There, Joe had his first encounter with a monocled cobra: It was a four-foot-long beauty, and he had to have it.

Crother liked the snake too, but he was supposed to be exercising adult supervision. He said, "You're crazy! You can't just buy a cobra. You have to get Savage's permission." Joe went to call their professor and came back to the hot room beaming, saying that Savage had given him the go-ahead. Joe also purchased his first professional snake-handling stick that day; before, he had just used his hands. He triumphantly returned to campus with the monocled cobra in a specimen bag, intending to put it in a forty-gallon vivarium in the lab.

Until then, Joe had mostly handled pit vipers such as rattlesnakes and copperheads, which don't like to fall, and will usually balance calmly on a stick when you pick them up. Cobras, however, are more aggressive predators, and much quicker than vipers. Back at the lab, when Joe reached into the bag with his new snake stick and brought out the cobra, it immediately leaped away, caromed off Joe's thigh, and escaped. (One version of the story features a hysterical female undergraduate who at this point in the narrative jumps onto a lab table and starts shrieking, but according to Brian Crother that's an embellishment—perhaps Joe's own.)

Jay Savage arrived at his lab to find two of his most promising graduate students trying to catch an agitated cobra, with other students coming in to watch the spectacle. It was a medical disaster—not to mention a lawsuit—waiting to happen. "We were all sweating bullets," Crother recalled. "At one point Joe said, 'You know, Jay, if you're scared, you can leave and I'll take care of it.' I said, 'Man, are you *crazy*?' He was talking to one of the most famous herpetologists in the world." If Joe was trying to be funny, it didn't go over well; Savage glared at him furiously and didn't say a word.

Eventually they caught the snake and got it safely back into the bag. Savage was livid: Apparently, Joe had told him on the telephone that it was a baby, not a fully grown cobra. The professor scornfully rejected Joe's explanation that, as monocled cobras go, four feet was on the small side. Characteristically, Joe's letter home describing the incident was both unrepentant and slightly disingenuous: "I bought a cobra which I brought back with the intent of keeping it in my office. However, my major professor walked in shortly after I had placed it in an aquarium, and told me to get rid of it. His only reason was that it might kill somebody—a pretty feeble reason for making me get rid of such a fine creature." Joe reported that he took it back to the shop and exchanged it for a "very cute" pair of baby crocodiles. "I have a room full of reptiles now—I've decided that this is enough, and I don't plan to acquire any more."

After the furor died down, Joe got a part-time job at Zoological Imports.

The cobra's aggressive posture when aroused—the head and much of the body erect, hood flared, the deep-forked tongue flickering—is the most widely identifiable snake image in the world. Cobra legend goes back to antiquity: The asp that Cleopatra used to kill herself was an Egyptian cobra. A rich tradition of folk natural history surrounds the cobra: It hypnotizes its victims before striking (actually, it doesn't—"frozen by fear" would be a more scientific description of the snake's effect on its prey); cobras can be charmed and made docile by music (they're deaf—in the familiar image, the snake charmer guides the snake's movements by swaying his head, not with the music of his flute). Racing champion and automotive designer Carroll Shelby called his legendary racecar the Cobra after the name appeared to him in a dream.

Yet when Joe Slowinski entered graduate school, cobras were drastically understudied. The situation has since improved, but there is still no agreement about what exactly a cobra is. Most snakes called by the name belong to the genus Naja, but that omits the longest and most famous species, the king cobra (*Ophiophagus hannah*), and the smallest, the ringhal of South Africa (*Hemachatus haemachatus*), which is just over a meter long. "Cobra" is a term of convenience for large elapids that can expand the neck ribs to form a hood, but there is no firm scientific definition.

Cobras inhabit a great arc around the Indian Ocean, from South Africa to Egypt, across South Asia, and throughout the Malay peninsula and archipelago, exhibiting a diversity remarkable even by serpent standards. The ringhals are viviparous, giving birth to as many as sixty live young at once; the oviparous king cobra builds a nest—a mass of dirt and forest litter two feet across—for its eggs, which are diligently guarded by both parent snakes until they hatch. The most common, widespread species is *Naja naja*, the Indian spectacled cobra, which is found throughout India and Southeast Asia. The villains in Rudyard Kipling's classic tale in *The Jungle Book* about the valiant mongoose Rikki-tikki are a pair of *Naja naja*. In the first encounter between the mongoose, the cobra's only terrestrial predator (besides *Homo sapiens*), and the male cobra, Nag, Kipling gives a precise, scientific description of *Naja naja*, evoking the snake's fearsome aspect:

> From the thick grass at the foot of the bush there came a low hiss—a horrid cold sound that made Rikki-tikki jump back two clear feet. Then inch by inch out of the grass rose up the head and spread hood of Nag, the big black cobra, and he was five feet long from tongue to tail. When he had lifted one-third of himself clear of the ground, he stayed balancing to and fro exactly as a dandelion-tuft balances in the wind, and he looked at Rikki-tikki with the wicked snake's eyes that never change their expression, whatever the snake may be thinking of.
>
> "*I* am Nag. The great god Brahm put his mark upon all our people when the first cobra spread his hood to keep the sun off Brahm as he slept. Look, and be afraid!"
>
> He spread out his hood more than ever, and Rikki-tikki saw the spectacle-mark on the back of it that looks exactly like the eye part of a hook-and-eye fastening. He was afraid for the minute; but it is impossible for a mongoose to stay frightened for any length of time, and though Rikki-tikki had never met a live cobra before, his mother had fed him on dead ones, and he knew that all a grown mongoose's business in life was to fight and eat snakes.

The curious spectacle mark on the back of the snake's hood, which resembles a nineteenth-century pince-nez rendered by an elegant squiggle of classical Chinese calligraphy, has generated ingenious explanations. Modern evolutionary theory inclines toward a defensive adaptation: The pair of vivid, eyelike marks, black outlined in white, might have evolved because they heighten the cobra's frightening appearance when it spreads its hood; they may even create the illusion that the snake is staring down its adversary when it retreats. Kipling gives the Hindu version of the myth; the Burmese believe that the very similar hood marking on the monocled cobra came from the kiss of the Buddha, given in gratitude after the snake shielded him from the sun. Both species exist in abundance, raising the fascinating possibility that human veneration of species with such enigmatic markings might have favored their survival because they were protected from predation by pious human hunters.

By the end of Joe's first term, he and Crother had become best friends, but there was little evidence that the older student was exerting much of a restraining influence. Just before Christmas, Joe tried to ship a diamondback rattlesnake by bus to Stan Rasmussen, who by this time was studying law at the University of Colorado. Joe taped up the snake's rattle to silence it, dropped the animal into a cotton pillowcase, and shipped it off in a cardboard box. In Omaha, when baggage handlers opened the bus's cargo bay, they found a torn box and a wriggling pillowcase.

The bus line called in the local Humane Society to take possession of the snake. A story in the *Omaha World-Herald* (TWITCHY PASSENGER RATTLES GREYHOUND) reported that the Humane Society director had turned over the names of the shipper and the intended recipient to an official of the Fish and Wildlife Service "to determine whether any federal laws were broken by the shipment." The return address on the box was DEPARTMENT OF BIOLOGY, UNIVERSITY OF MIAMI. When the federal authorities phoned, Jay Savage took the call. Afterward, he told Joe, "If they prosecute you, you're out of here."

Whatever he might have said to the police, Savage was surely rooting for Joe: He had already begun to develop a profound respect for Joe's mind that endured long after Joe's studies with him had ended. "He was a deep thinker on different subjects," Savage would later say. "That's unusual in a young scientist." To everyone's relief, the authorities decided not to bring charges, which probably saved Joe's career.

Even if his judgment was still green, Joe grew up physically in graduate school. He filled out his five-foot-eleven frame, and his hair lost its flossy golden luster, darkening to auburn. He began to look like a grown-up. Inseparable outside the lab, he and Brian Crother became regulars at several bars in Miami. Their favorite place to play pool was a country-western sports bar called the Stampede. Crother recalled, "We would arrive and see a couple of pickup trucks in the parking lot, and Joe would say, 'The 'Pede is packed again.' "

Despite his penchant for honky-tonks and lax safety standards in the transport of venomous snakes, Joe was soon off to an impressive start as a biologist, specializing in the challenging field of evolutionary theory. Some of his best papers germinated on the biology department's porch, known as the Veranda, where Jay Savage and his students gathered at sundown to drink a few beers and unwind. The conversation was mostly about sports, movies, and academic gossip—but occasionally some real science got done. One afternoon Joe and Craig Guyer, a student of Jay Savage's who was close to getting his Ph.D., realized that they had both recently read and been intrigued by an essay with the emphatic title "There Have Been No Statistical Tests of Cladistic Biogeographical Hypotheses!" by Daniel Simberloff and other scholars.

Simberloff and his coauthors pointed out that certain types of statistical analysis routinely used by scientists in other fields had never been taken up by biologists to explain diversity in the evolution of species. Their paper was focused on biogeography, which studies why some species evolve (and others become extinct) in certain areas and not in others. The title of the essay laid down a challenge, which Joe and Guyer gamely took up: The two men set out to fill the void Simberloff had described—on a grand scale.

Their response, a couple of years in the gestating and drafting and revising, was published in *The American Naturalist* under the title "Testing the Stochasticity of Patterns of Organismal Diversity: An Improved Null Model." They developed a series of equations known as a null model, which was intended to eliminate the possibility that chance (stochasticity) alone could explain patterns of diversity. Their article wasn't concerned with any particular species; it could apply to amoebas or whales as well as to cobras. It was an important conceptual breakthrough and has often been cited; Joe was still three years away from being awarded his Ph.D. when the paper was published.

In the fall of 1988, a big, blond fellow from Indiana enrolled at the herps lab in Coral Gables. Brady Barr was a former high-school teacher with an enthusiasm for alligators. Joe took Barr under his wing in much the same way that Brian Crother had looked after him. The two herpers spent many a night in the Everglades chasing snakes and alligators, and many more hanging out in bars in Miami. Barr would later say in his signature country drawl, "Joe was a hard-drinkin', fast-smokin', fast-talkin' kind of guy. He'd stay up damn near the whole night—and then he would be the first one up in the morning."

Yet in Barr's recollection, Joe was above all a brilliant scientist; Barr called him "the smartest guy I ever met." Barr was amazed by Joe's dedication as a journal-keeper. In the field, no matter how wet, exhausted, and hungry he was at the end of the day, Joe always sat down and updated his journal, noting the specimens he had collected and under what conditions, and embellishing his lists with gossipy accounts of his doings throughout the day and opinions on everything from college basketball to rock 'n' roll.

Joe was a less kindly big brother than Brian Crother had been: He teased Barr mercilessly. When they went camping, Barr liked to go to bed early, but Joe would stay up drinking beer and gabbing as long as there was someone to join him. After Barr had retired, Joe, when he finished a beer, would throw the empty can at Barr's tent. The length of time it took Joe to drink a can of beer, it seemed, was exactly how long it took Brady Barr to drop off to sleep again; then Joe would lob another strike. Joe's inner twelve-year-old, his companion for life, never tired of this game. The best part was hearing Barr's sleepy, empty threat: "Don't make me come out there."

Herpetology is a chimera among academic disciplines, uniting the study of two groups of animals that aren't very closely related, from an evolutionary perspective. Taxonomically, reptiles are much closer to birds than they are to amphibians. Strictly speaking, birds *are* reptiles: In most modern taxonomies, they share the class Archosauria with the crocodilians and dinosaurs.

Nonetheless, the study of amphibians and that of reptiles have always been joined, following the practice of classical Greek science to lump together all creeping (*herpetos*) animals. "In part, it is historical inertia—some would say tradition," as the leading college textbook, *Herpetology: An Introductory Biology of Amphibians and Reptiles*, puts it, with a touch of acerbity. (The senior author of the textbook is George Zug, Joe's codirector of the Myanmar Herpetological Survey.) Although the union is firmly established at educational and research institutions, there are now some fissures, suggestive of a possible split. However, if such a change ever takes place, it will be long in coming, for biologists studying the two groups share many common concerns. Snakes and frogs often coexist in nature; being eaten by a snake is one of the leading causes of premature frog death.

Perhaps in part because of this lopsided balance of power between the animals they study, herpetologists who specialize in reptiles have always possessed more prestige in their field than those who study amphibians: No frog can compete with the king cobra or the black mamba in sex appeal. Yet over the past twenty years, scientists in many fields have focused intently on amphibians, which are declining at a cataclysmic rate—a phenomenon that may have profound implications for all living things. Particularly in the Western Hemisphere and Australia, herpetologists specializing in amphibians are exhausting their hoard of gloomy superlatives to describe frog die-offs, steep declines in and even the total disappearance of populations, which have no clear explanation.

British herpetologist Tim Halliday, the international director of the Declining Amphibian Populations Task Force, based at the Open University in Milton Keynes, England, confirmed that mass declines in amphibian populations are worsening. He cited a global assessment published in 2005, which concluded that a third of the world's amphibian species are threatened with extinction. "The most bizarre aspect of this," said Halliday, "is that in order to make a real assessment, we have to determine what's out there, but species are disappearing so rapidly, we can't do that." Many reasons for the phenomenon have been proposed, such as rising levels of ultraviolet-B radiation in sunlight, pollution, introduced predators, and overharvesting for food and medicine. Recent research has described a plague of the chytrid fungus, *Batrachochytrium dendrobatidis*, which kills the amphibians by rotting their skin, an essential organ of respiration.

Many scientists use the familiar analogy of the canary in the coal mine: Whatever is that bad for frogs can't be good for people. Yet Tim Halliday said, "Personally, I don't buy the canary-in-the-coal-mine analogy. I think what amphibians are telling us is that severe problems exist in their own habitats." He pointed out that while amphibians are declining at a faster rate than most taxa, just about every class of animal life is falling off, everywhere—except human beings. Halliday sees the severe declines in amphibian populations as a symptom of a larger problem: the catastrophic degradation of freshwater habitats worldwide.

Guinevere Wogan, a herpetologist at Cal Academy who accompanied Joe Slowinski on many expeditions, including the one to Upper Burma in 2001, said, "In Myanmar we have no evidence of large-scale die-offs of amphibian populations. Actually, there is no evidence that this phenomenon is occurring anywhere in Southeast Asia." However, she added, that may simply be a reflection of the spottiness of observations in the region. On her next trip to Burma, she plans to swab frogs in order to determine if chytrid is present.

When Joe arrived at the University of Miami, biologists were deeply involved in ecology and efforts to preserve biodiversity, but their main mission remained the same as it had been since Darwin rationalized the basic principles of evolutionary biology: to complete the census of life on Earth and describe as precisely as possible the relationships among species. The intellectual core of modern biology was economically expressed in the title of Theodosius Dobzhansky's essay "Nothing in Biology Makes Sense Except in the Light of Evolution." Jay Savage added the corollary that nothing in evolution makes sense except in the light of phylogeny. Phylogeny is another word for systematics, the study of the evolutionary relationships among species, which strives to establish a detailed, verifiable tree of life that accounts for the vast complexity of plant and animal species.

In 1984, the means of describing those evolutionary relationships was expanding exponentially. When the technique of gene-sequencing was first discovered, in the early 1960s, big hunks of liver and heart were required to conduct the analysis, so any widespread application was impractical. Yet by the time Joe Slowinski began his studies in Miami, polymerase chain reaction technology had made it possible to sequence genes using very little tissue. A few cells were all that were needed to create a genome, a detailed, digital map of a species' genetic code. These technological advances made it far easier to determine, to a fine degree of precision, how much genetic material is shared by tissue collected from different animal specimens.

Studies of this sort reach their conclusions by manipulating masses of DNA-sequence data using complex statistical analysis—and that requires a strong working knowledge of higher mathematics, something Joe had neglected in his studies at the University of Kansas. If he was to be a player in the ever more abstruse field of contemporary herpetology, he needed to do some cramming. In Miami, in addition to his course work, Joe mastered calculus and scoured the literature for the latest statistical models. He proved to have a remarkable aptitude for higher math; on his own initiative, he quickly ascended into ever more rarefied realms of symbolic thought.

Systematics sometimes resembles an art more than a science. It is an intellectual world apart from field biology, which dominated Joe's mature career as he led expeditions into the jungle of Burma and other remote wilderness localities. Systematics is a purely theoretical discipline that can never entirely resolve the complex issues at its heart. It is the equivalent in biology of Zeno's paradoxical arrow, which is always getting closer to its target but never arrives.

What keeps scientists battering away? Jay Savage, when asked why scientists expend so much thought to describing these evolutionary relationships, replied, "It makes us feel good."

Joe himself eventually developed a sardonic attitude toward the theoretical pursuit that captivated his scientific imagination. A few years after he received his Ph.D., when he was teaching in Louisiana, he told a first-year graduate student, "If people knew what we were using the taxpayers' money for, to find out how one species of snake is a tiny bit different from another, they would come after us with torches and clubs like the mob in *Frankenstein*, to get their money back."

Central American Coral Snake

Micrurus nigrocinctus

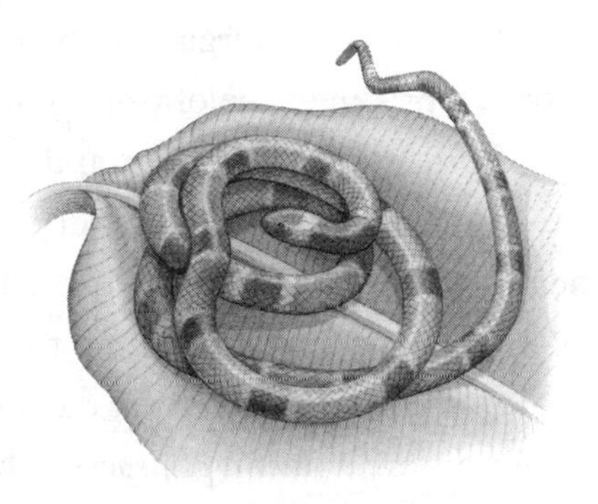

The Central American coral snake is one of the handsomest denizens of the rain forest, with bright crimson and narrower black bands, demarcated by delicate ocher rings. Like all elapids, it uses a pair of small fangs, which are fixed in the front of the top jaw, to deliver venom. Its neurotoxic venom is potent and complex, but may be slow to take effect; therefore the snake tends to hold fast to its victim when biting. Central American legend holds that the snake stings with its tail, because of its dramatic defensive posture: When it is threatened, the snake buries its head in its coiled body and extends the tail in the air, slashing it back and forth. Like most coral snakes, it is primarily ophiophagous (that is, it feeds on other species of snakes), but it will also consume lizards and, infrequently, small rodents. The species is active at early evening and nighttime. In his study of the snakes of Costa Rica, Jay Savage writes, "A common snake in all lowland and premontane forest areas, often found under debris, in pastures, on coffee fincas, and in vacant lots and gardens in urban areas."

Except for family vacations in Mexico during his childhood and a couple of daylight sorties across the Rio Grande with Stan Rasmussen, Joe had not journeyed outside the United States since his childhood visit to Kyoto, now a distant memory. So when the opportunity arose for a summer of research at the Organization for Tropical Studies in Costa Rica, he seized it. Unlike the woods of Kansas and the marshlands of the Everglades, the Costa Rican rain forest was an alien world, a pristine zone of tropical life. The diversity of fauna was dazzling, from gaudy butterflies and parrots to the majestic jaguar. The forest's deadliest inhabitant, very much to Joe Slowinski's taste, was the bushmaster (*Lachesis muta*), the largest venomous snake in the New World, attaining a length of up to twelve feet—an animal so rare as to be almost mythical. Nearly as dangerous, and far more numerous, was the fer-de-lance (*Bothrops asper*). Like the bushmaster, the fer-de-lance is a pit viper with enormous fangs and ample reservoirs of its flesh-devouring venom.

The summer term began at Tortuguero National Park, a tract of rain forest on the Caribbean coast so remote that it's inaccessible by land: Savage and his students went in on a barge and came out on a plane. Tortuguero supported thriving populations of caimans, manatees, sloths, river otters, three species of monkeys, and the eight-hundred-pound green turtle (*Chelonia mydas*) that gives Tortuguero its name; the park's beaches are the turtles' principal nesting site in the Caribbean Sea.

The summer in Costa Rica was Joe's first chance to engage in a prolonged, intensive period of fieldwork—like his high-school mammoth dig, but with himself now in the role of scientist, practicing his vocation. Cofounded twenty years earlier by Jay Savage, the Organization for Tropical Studies was essentially haute science camp for graduate students: an opportunity for them to get their feet muddy doing some primary, publishable science.

Early in the morning, the students would go into the field and make a round of observations. When they got back to camp, they would define a problem. For example: As you walk through the rain forest, you will notice that the leaves of some plants are eaten by insects more than those of others; you will further observe that those less eaten by insects have epiphytes such as ferns and orchids attached to their boughs. Does it therefore follow that the presence of epiphytes inhibits herbivory? After formulating the scientific problem, the students would brainstorm and create an experimental method to study it. The rest of the day was spent collecting data and trying to find a solution, which they would present in the evening. Brian Crother said, "It was a brilliant way of getting the students to think creatively, like 'real' scientists."

The Organization for Tropical Studies has produced hundreds of scholarly papers since it began. Joe Slowinski's first scientific publication, a very brief article coauthored with Crother and a graduate student from Duke named John Fauth, grew out of one of those observation-problem-solution exercises. The title, "Diel Differences in Leaf-Litter Abundances of Several Species of Reptiles and Amphibians in an Abandoned Cacao Grove in Costa Rica," was almost as long as the paper, and told pretty much everything it had to say. ("Diel" is scientific jargon meaning daily, descriptive of a twenty-four-hour cycle.) The authors went into a deserted cacao plantation and made a census of five species of lizards and frogs, day and night, noting whether they were scurrying on top of or underneath the dead leaves. The article was published in the *Revista de Biología Tropical*, a journal of the University of Costa Rica. It was about as humble as a scientific paper can be, but it got the job done.

At Tortuguero, Joe saw his first coral snake in the wild—a fateful encounter, for he would soon decide to write his Ph.D. thesis about coral snakes. He and Crother were hiking in the lowland rain forest one night when a Central American coral snake flashed across the trail. Joe was instantly aroused and leaped after it. He managed to pin it with his snake stick, but the strong, lithe elapid wriggled right out from under it and shot off into the forest. Brian Crother said, "What are the three worst words for a field herpetologist? 'It got away.'" After that experience, Joe threw aside caution—and his snake stick—and caught snakes with his hands, as he had done when he was chasing copperheads in Kansas.

As a group, coral snakes have a bad reputation even among herpetologists, who can usually find something good to say about any snake. Field guides always describe them as secretive; herpetologist Harry Greene adds to that "supple, surreal, treacherous, and unpredictable." American coral snakes have a bizarrely lopsided phylogenetic family tree: In 1995, Joe would publish a landmark paper establishing two genera of New World coral snakes. One of them comprises but a single species, *Micruroides euryxanthus*, the Sonoran or Arizona coral snake; the other genus, Micrurus, accommodates a remarkable sixty species, making it one of the most diverse taxa of serpents.

Yet all the corals share many distinctive traits. Like the king cobra, they mainly hunt and feed on other, smaller snakes (though the harlequin coral snake, *Micrurus fulvius*, the most familiar venomous serpent in the eastern United States, is known to consume prey with a body weight that far exceeds its own). Coral snakes have acquired some dramatic, even bizarre, defensive adaptations: The male of the Southern coral snake (*Micrurus frontalis*) pumps its pair of penises quickly in and out of sight to intimidate predators. If *Micruroides euryxanthus* is threatened, it buries its head in its coiled body and waves its tail in the air as it forcefully everts its anal lining, ejecting the contents with a popping sound that can be heard from a meter away. This snake survives on a highly specialized diet of blind and black-headed snakes.

The evolutionary success of the New World coral snakes has inspired other, nonvenomous snakes to mimic them as a defensive adaptation—the "dead ringer" strategy exemplified by the Dinodon and the many-banded krait. One of the best-known examples of the phenomenon is the nonvenomous scarlet king snake, *Lampropeltis triangulum*, which mimics the harlequin coral snake: It's the same size as the venomous elapid and has very similar coloration, with alternating bands of black, yellow, and red. However, it is less precise in its mimicry than the Dinodon is in its imitation of the krait: The colored bands are in a different order.

The difference between the two species has been drilled into the brains of millions of schoolchildren in the American South, where the harlequin coral snake thrives. A scrap of doggerel about the sequence of color bands helps them to distinguish between the two: "Red touches yellow, kill a fellow. Red touches black, venom lack." This adaptation evolved around twenty million years ago, in the same way that the Dinodon's mimicry of the krait developed: Birds of prey and other animals that hunted snakes avoided those ancestors of the scarlet king snake that resembled the deadly coral snake (paying no heed to the different order of the colored bands), resulting in the gradual dominance of the harlequin-like markings.

It's a remarkably widespread evolutionary tactic: More than one hundred New World colubrids mimic coral snakes. The adaptation is particularly common in serpents, but a superficial resemblance to coral snakes gives an edge for survival to animals that otherwise have nothing in common with them. Scientists believe that in northwestern Costa Rica, where *Micrurus nigrocinctus* is the only coral snake, creatures as utterly dissimilar as the caterpillar of the frangipani hawkmoth (*Pseudosphinx tetrio*) and the ornate wood turtle (*Rhinoclemmys pulcherrima*) have evolved a resemblance to the snake, which repels predators.

When the summer term at the Organization for Tropical Studies ended, Joe was loath to return to Kansas City, where he would be pressured into taking a job mowing lawns or painting houses until the fall semester started. He stayed on in Costa Rica to help Brian Crother collect the snake specimens he needed to complete his Ph.D. dissertation. With another friend, Joe climbed Cerro Chirripó, the country's highest peak. On that trip Joe collected a very rare colubrid species, *Urotheca myersi*; his was only the second known specimen.

Joe's experiences in Costa Rica awakened within him a lifelong wanderlust—the passionate love of desolate places that field herpetologists share with religious mystics and soldiers of fortune. He continued to explore Florida, particularly the Everglades, with its rich crocodilian fauna; yet as his graduate studies progressed, he came to view his profession partly as a way to see the world beyond the United States.

He began with the Caribbean, North America's tropical backyard. One summer, he and Crother spent ten fruitless days in Jamaica searching for a species of West Indian racer that they later decided was probably extinct. On another trip, they spent a couple of weeks on an island called Frazer's Hog Cay, in the Berry Islands chain of the Bahamas. "It was basically a rock in the middle of the sea," said Crother. "You could walk around it in a day." They camped on the beach, idling and drinking beer when they weren't seeking another rare species of racer, *Alsophis vudii*. Joe got Crother into trouble on that trip: He smuggled out a boa constrictor without a permit, and he didn't tell his friend. The first Crother knew of the snake's existence was when the authorities discovered it and started asking hard questions he couldn't answer.

On three visits to South America, Joe pushed into ever more remote terrain. The first trip came at the end of the summer of 1987, when he took a job as a shipboard naturalist aboard a jungle-river cruise in Peru. After the job ended, he flew to São Paulo to visit the Instituto Butantan. There, he exhibited what was becoming a habitual ruthlessness when an opportunity to collect herps presented itself. When he went to visit the director, who had a spectacular collection of snakes, he encountered a linguistic obstacle: Joe spoke passable Spanish, but the director did not. They were at loggerheads, speaking scraps of Spanish, Portuguese, and English to each other as Joe tried to communicate what he hoped to do with her collection. Finally, the director gave up, but managed to convey to him that she had another appointment. Joe understood her to say, "Do whatever you have to do, and I'll be back in an hour." She left him in the hands of her deputy.

In those days, zoologists who wanted to collect samples that they could later use for biochemical analysis had to use portable refrigerators cooled with liquid nitrogen, so they could freeze the specimens. When Joe visited São Paulo, he had his nitrogen refrigerator at the ready: As soon as the institute director left the room, with the aid (or at least the acquiescence) of her assistant, Joe began loading up his refrigerator with venomous snakes as fast as he could stuff them into it. The snakes died quickly in the freezer. When the director returned, she politely asked her American guest if he had gotten what he needed. Joe answered in the affirmative.

Then she looked around and asked, "Where are my snakes?"

Despite the language barrier, even Joe understood her question: He opened the refrigerator and showed her. As he told the tale to his father, one of the last snakes to be deposited in the freezer wasn't quite dead, which Joe found embarrassing; its tail stuck up and wiggled slightly. Stretching a point, as he had done with the "baby cobra" that got loose in Jay Savage's lab, Joe would later plead innocence on the grounds that the director had told him to do what he needed to do, so that's what he did.

His next trip to Peru was an adventure with his sister. After their extraordinarily close relationship in childhood, Joe and Rachel had followed the normal course of adolescence and declared their independence of each other. But in their twenties, as they both found a calling in life, Joe in herpetology and Rachel as a photographer, they again became very close. The trip to Peru was a lark more than an expedition: One day Rachel announced that she wanted to go to Peru, and Joe said, "Me too," declaring that he wanted "to go look for lost Indian tribes."

Rachel went ahead, and Joe promised to meet her there. They didn't have a definite rendezvous, but Joe habitually traveled with the hopeful expectation that everything would sort itself out. Rachel recalled, "He said, 'Meet me in Lima on the tenth,' and then in Lima on the tenth I found a note saying, 'Meet me in Cuzco.' There was no plan at all—it was really stupid. We were just out looking for lost Indian tribes." They did manage to find each other, but soon discovered that it was a difficult time to be in Peru; El Sendero Luminoso, the Shining Path rebel group, was active, so travel inside the country was irregular and sometimes dangerous. "The Sendoso had organized a nationwide strike," said Rachel. "We were just walking through the countryside one day, and there was a big demonstration. We were sprayed with tear gas and got stranded in this little village for a week." Even more worrisome, the Shining Path had recently killed some American Peace Corps volunteers working in the mountains.

After they had been on the road a few weeks, riding in the back of a gold miner's pickup truck, Joe's passport and money were stolen, so he had to make his way back to Lima. He told Rachel to meet him at Puerto Maldonado, a town in the middle of the Amazon jungle. Rachel had trouble booking a flight and arrived for the rendezvous a few days late aboard a propeller plane that landed in a cornfield. Joe was nowhere to be found. As she sat on a park bench in the plaza, pondering what to do next, she started talking to a stranger who told her that he had met a blond-haired foreigner a few days before. He couldn't remember the man's name; he only recalled that when he saw a large beetle, he caught it and put it in his pocket. The man on the park bench said he thought the *norteamericano* was staying at a lodge downriver. "So I took a boat to the lodge," said Rachel, "and Joe was waiting. The first thing he said was, 'About time you showed up.' He knew I would find him." They never found the lost Indian tribes, but they spent a few days there, exploring a tranquil lake full of caimans.

Not all of Joe's early travels were so rich with adventure. In 1989, he went on his first (and only) trip to Europe, to attend the First World Congress of Herpetology in Canterbury, England. After the congress, Joe went to Paris for a few days. He signed up for a guided tour, which shuttled the tourists from one famous sight to the next. Joe hated it: He was more comfortable in the back of a pickup truck, careering through the wilderness of Peru, than being herded through cathedrals. Anyway, Europe was a poor place to look for snakes. He did make one important connection on the trip: At the hotel bar in Canterbury, he met the German herpetologist Wolfgang Wüster, who would later become a key collaborator.

Joe never told anyone that the coral snake that got away at Tortuguero National Park was responsible for his decision to write his dissertation about the phylogeny of New World coral snakes, but he made up his mind soon after his return. He started reading and comparing the existing literature on coral snakes, such as it was, and realized that no one really understood their evolutionary history. Jay Savage liked the idea: It was a big, important subject with plenty of controversy and thus perfect for a doctoral thesis. Savage was curious to see what Joe would come up with.

Joe set about acquiring specimens: He would need as many as he could get to carry out the physical measurements and biochemical research he would need for his dissertation. His quest for coral snakes struck an unexpected bonanza at the All-Florida Herpetology Conference that year.

The annual conference in Gainesville is one of the most popular such symposiums in the country, where leading scholars gather to exchange ideas and name cards. The event is also something of a jamboree, attracting amateurs and collectors for workshops on the care and breeding of herps, and auctions of prize specimens. Jay Savage was a fixture at the conference and encouraged his students to come. At a barbecue banquet on the last night, Joe met a pianist named Tommy who collected coral snakes; he told Joe his basement at home was stacked with vivariums full of them. Joe asked him if he would save any snakes that died, so that he could use them for comparative purposes. Tommy promised he would.

Six months later, he called Joe to tell him that he had a sackful of dead coral snakes for him to come and collect. He told Joe he was working at a resort on the beach, playing piano, and gave him the address. When Joe arrived, he reported to the security desk at the entrance and started to walk through. A guard stopped him, saying, "Wait a minute, you can't go in like that. You have to take off your clothes." Joe was stunned. The guard explained: "Everyone inside is naked. This is a nudist colony."

Joe was on the sharply pointed horns of a dilemma: He desperately wanted those snakes, but the prospect of public nudity appalled the middle-class Kansas City boy in him. When he regained his composure, he tried to talk the guard around. Eventually, they reached a compromise: Joe could go in wearing a towel. So he stripped down, wrapped a towel around his waist, and walked into the resort. Joe didn't know which way to look: People were swimming in the pool and playing tennis, sunbathing and reading, all in the nude.

Joe found Tommy, quite naked, playing piano in the ballroom, and urgently asked him about the snakes. "As soon as I finish this set," he answered. Joe sat on a little chair by the piano, blushing crimson as the guests stared at him sourly, disapproving of his towel. Finally, Tommy finished his set and led Joe away to collect the coral snakes. When Jay Savage tells the story, it concludes with a wicked, gap-toothed grin: "Joe wanted those snakes, and he did what he had to do to get them."

RON SLOWINSKI

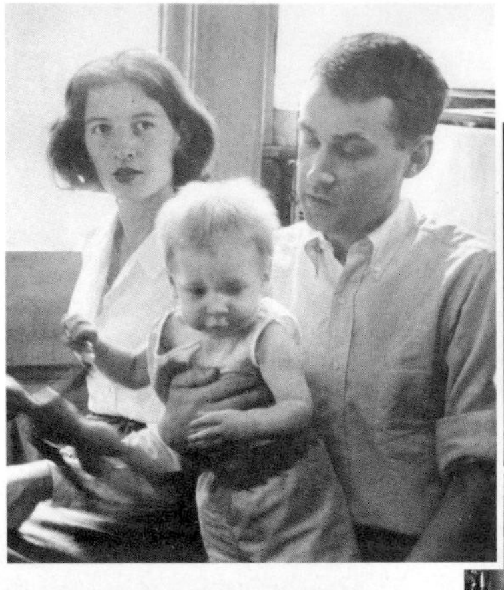

A peripatetic childhood *(clockwise from upper left):* Martha and Ron Slowinski with their firstborn, Joe, in New York, in 1963; Martha with Joe and his sister, Rachel, at the family's summer cabin in Door County, Wisconsin; Joe, Rachel, and Martha outside a Buddhist temple in Japan; Joe and Rachel spelunking in New Mexico. ∞

RCN SLOWINSKI

Joe and Rachel with butterfly net and cricket cage, in Kyoto. ∞

RON SLOWINSKI

In New Mexico, Joe and Rachel met a Taos war chief named Sunhawk, who showed them a herd of wild bison. ∞

Etching of an iguana by Joe Slowinski, age 14. ∞

STANLEY RASMUSSEN

On the banks of the Kaw River, Joe displays a recently unearthed bison skull.

RON SLOWINSKI

At the University of Kansas, in 1982, Joe wrangles a timber rattlesnake (*Crotalus horridus*).

RON SLOWINSKI

In 1979, Joe was a volunteer at a fossil dig, the Mammoth Graveyard, in South Dakota. ∞

Father and son tour the Everglades. ∞

BRIAN CROTHER

Joe Slowinski and Brian Crother partying after a golf tournament in Austin, Texas. ∞

ASHLEIGH B. SMYTHE

Joe and Ashleigh Smythe, on a camping trip in Big Bend. ∞

RACHEL SLOWINSKI

Beside a towering kapok tree in Peru. ∞

ALAN REBERTUS

Joe Slowinski in Costa Rica in 1985, with a boa constrictor.

Pygmy Rattlesnake

Sistrurus miliarius

The pygmy rattlesnake, which rarely exceeds two feet in length, lives in the pine woods and wetlands of the southeastern United States. Sometimes called "ground rattlers," these diminutive snakes are so well camouflaged that they are rarely seen. When coiled on the forest floor, they are the size of a pinecone. They have a small, delicate rattle that is not much wider than the tip of the tail. When they vibrate it, it sounds like an insect buzzing; some people have difficulty hearing the rattle. Pygmy rattlesnakes are passive hunters: They sit and wait for prey to come their way. Their preferred diet includes mice, frogs, lizards, and other snakes; they will also eat insects, spiders, centipedes, and nesting birds. The snake has a reputation for being very aggressive and apt to strike if disturbed. It injects small quantities of venom when it bites. Although there is no record of a human dying from the bite of a pygmy rattlesnake, it's extremely painful and can result in the loss of a digit.

After Joe successfully defended his dissertation on coral snake phylogeny, and the long-coveted honorific "Doctor" was finally prefixed to his name, he confronted the vaguely terrifying question: Now what? The tide seemed to be moving against him: It was a terrible time to be looking for a job in herpetology. Since the boom in postgraduate studies during the Vietnam War era, when they offered educational deferments from the draft, Ph.D.'s had become almost a common commodity. There weren't enough good positions to go around, even for highly qualified candidates.

Brian Crother was one of the lucky ones, landing a tenure-track post at Southeastern Louisiana University soon after he received his doctorate; most scientists like Joe with freshly minted Ph.D.'s turned to postdocs. In those days, postdocs were essentially workfare for the overeducated, stipends that permitted scholars with doctorates who couldn't find a full-time affiliation with a research institution to continue an investigation begun in graduate school or to undertake a new line of research—and to put pizza on the table without having to get a job delivering it.

The year after Joe completed his studies at Miami, George Zug hired him for a postdoc position at the National Museum of Natural History, a part of the Smithsonian Institution in Washington DC. Zug had received his Ph.D. from the University of Michigan in 1969 and was immediately hired as a curator at the NMNH—a position he held until his retirement in 2007. He said, "My generation walked out of grad school into good jobs. Today, a young scientist fresh out of grad school has a better chance of getting struck by lightning than getting a job as a curator." By the twenty-first century, postdocs had become a part of the journey: Most advertisements for prestigious research appointments today routinely specify, "Postdoc experience required."

Joe's sojourn in the capital had a temporary feel from the start. He lived in a basement studio in a townhouse in the Capitol Hill district. Setting the pattern for his future residences, it was sparsely furnished, with a coffee table and a futon on the floor, his parents' paintings and a deer skull with antlers on the walls, and little more. He drove a junky

Datsun Z-series convertible, which he ironically called his "chick magnet."

Yet he formed some enduring friendships there, principally with Zug and Roy McDiarmid, the co-supervisors of his project. Among his peers, Joe quickly found a friend in a herpetologist named Kevin de Queiroz. Although de Queiroz was slightly older, Joe was his academic "uncle": Herpetologists have established an intellectual genealogy for themselves by likening academic supervisors to fathers. De Queiroz's Ph.D. supervisor, herpetologist David Wake, was a student of Jay Savage's, thus making Joe, another of Savage's progeny, Wake's "brother." Roy McDiarmid also studied under Jay Savage, one of the most prolific Ph.D. advisors of his generation, making him another of Joe's academic brothers.

Joe was attracted to the notion of herpetology as a craft, like a medieval guild: This fanciful academic lineage reinforced the concept of herpetology as an intellectual tribe. The congress that Joe had attended in Canterbury published a lavish commemorative book called *Contributions to the History of Herpetology*, by Kraig Adler, which contained a detailed genealogy of the field. Adler extended what has pretentiously been called the A lineage back to Louis Agassiz, the Swiss-born naturalist who founded the Museum of Comparative Zoology at Harvard; Agassiz was in turn the protégé of Alexander von Humboldt, the great early nineteenth-century scientific explorer of the New World.

Joe's postdoc marked a major shift from the New World to a concentration on the venomous snakes of Asia. He undertook the first-ever study of the systematics of Bungarus, the genus of the Elapidae that includes the kraits, based upon osteological and hemipenial data—in other words, the species' bones and male sex organs. Like all male snakes, kraits possess a pair of penises, called hemipenes ("half penises," from the traditional, erroneous belief that they combined to make a whole). The hemipenes inflate when inserted into the female; in some species, hemipenes are forked or spined, to ensure that they stay inside the female during what can be a long mating session—up to twenty-five hours for *Crotalus atrox*, the western diamondback rattlesnake.

It was a model postdoc. Most scholars who came to Zug created projects that were designed to go on forever—or at least until the applicant found a good job. "They never reached a conclusion," Zug said, "so they would have an excuse to seek an extension and get more money. Joe was the exception." He finished the data collection and wrote his paper, which was eventually published in the *Journal of Herpetology* in the allotted year.

In 1992, Joe moved to Baton Rouge for another postdoc, this time at Louisiana State University. He would live in the Bayou State for five years. The project, basically a continuation and expansion of the work he had done on Bungarus in Washington, was a study of the systematics of all the Elapidae. He chose LSU because he wanted to study with a biologist named David Good, who was an expert on DNA sequencing—in those days a vastly more complicated procedure than it is today.

Although much of Joe's time was devoted to the laborious tasks of collating data, measuring specimens in various ways, and interpreting the results with complex computer programs, his research lay near the heart of the ongoing, fundamental debate among biologists about how evolution works. Since his early years at Miami, Joe had read widely in the contentious literature of contemporary evolutionary biology. By the 1980s, straight neo-Darwinism—strict adherence to the doctrines of Charles Darwin—was becoming a minority viewpoint among biologists. Various hypotheses that relied on the basic tenets of the original system, modified by the observations of contemporary science, jostled noisily for support.

Joe gravitated toward phylogenetics, which was emerging as the new consensus view of evolution among biologists. Joe explained his understanding of phylogenetics in a book review published in the *Journal of the History of Science Society* in 2001. According to his review, the book, *In Search of Deep Time*, by Henry Gee, describes "the ascendance in paleontology during the 1970s and 1980s of phylogenetics, replacing the traditional storytelling involving ancestor-descendant sequences and adaptive scenarios." In phylogenetics, the emphasis is on finding relatedness among species. Joe made it clear in his review that he believed that this shift in the study of fossils reflected a broader change in biology as a whole. The molecular study of the elapids that Joe undertook at LSU was a classic exercise in phylogenetics.

Phylogenetics, in Joe's concise definition, is "the biological discipline that attempts to reconstruct the evolutionary relationships of organisms." To cope with the gaps in the fossil record, paleontologists before the rise of phylogenetics conceived of evolution in simple linear and narrative terms, expressed in the once-common panel illustrations "depicting the triumphant rise of modern humans from a succession of fossil hominids," in Joe's words, as though human evolution was primarily a matter of improving posture and personal grooming. Joe returned to a theme he had raised in his paper with Craig Guyer for *The American Naturalist*: Using deductive reasoning to explain adaptations, creating cause-and-effect story lines, can be very persuasive—and wrong.

In his review of Gee's book, Joe cited the traditional view of the evolution of birds, which held that certain unique avian features such as feathers and the hollow wishbone were adaptations that evolved to enable flight. This theory was debunked, Joe wrote, by "the phylogeny of some recently discovered fossils of birdlike dinosaurs (or dinosaurlike birds, if you prefer) from China, which revealed that the supposed adaptations for flight, including feathers and the wishbone, emerged long before flight." Phylogenetics threw out storytelling in favor of testable hypotheses that explained how species are related to each other in evolutionary terms, often depicted graphically in branching diagrams called phylogenies. Joe concluded on a visionary note: "The rise of phylogenetics has made possible a new universe of knowledge, a new way of understanding the history of life."

Opponents of modern biology, such as the advocates of "scientific creationism" and "intelligent design," have attempted to exploit disputes among evolutionary biologists to discredit the validity of Darwin's theory. However, since the publication of *On the Origin of Species*, it has been obvious to scientists that a hypothesis so far-reaching would have to be subjected to ongoing reinterpretation in order to keep it consistent with new information about terrestrial life. In the twentieth century, new technologies accelerated the pace of discovery in the life sciences to a hurtling speed. Yet the vast majority of contemporary biologists—approaching statistical unanimity—affirm the soundness of Darwin's basic, breakthrough premise: that higher life forms on Earth evolved from lower forms over long spans of geologic time.

Joe Slowinski detested creationism, and Louisiana was one of the places it took root. In the 1980s, when the movement was at its height, its proponents persuaded Arkansas and Louisiana to enact laws requiring that public-school curricula put the beliefs of creationism on a par with evolutionary biology. Both laws were struck down by the courts, but popular sentiment in favor of the faith-based system remained fervent in some quarters. In response to a letter to the *Advocate*, Baton Rouge's daily newspaper, by a supporter of creationism, Joe wrote an eloquent, if exasperated, rebuttal:

"As an evolutionary biologist, I find his letter frustrating and tiresome because it parrots the same discredited criticisms of evolution that modern creationists have apparently learned by rote. Evolution is a scientific theory which, like any other good theory, makes predictions that can be tested." Joe then systematically demolished the previous writer's arguments. He addressed the relative scarcity of transitional life forms in the fossil record with a herpetological example, complete with a scholarly citation:

> This statement reflects unbelievable ignorance. Ask any paleontologist: museums are filled with transitional fossils. In fact, one such form, an early Jurassic caecilian, was just described in the prestigious British journal Nature (1993. Vol. 365: pages 246–250). Caecilians are elongate, legless amphibians. The fossils reported are of a species of caecilian possessing the ancestral characteristic of legs, yet also possessing the advanced features of modern caecilians such as elongated body, tentacular fossa, etc.
>
> Sure, evolution is a theory. But the numerous predictions that it makes are strongly supported with evidence from every branch of science.
>
> JOSEPH B. SLOWINSKI, PH.D.
> Museum of Natural Science

It was surely the first time readers of the *Advocate* had been asked to ponder the phylogenetic significance of the tentacular fossa of the caecilian. (Tentacular fossae are pits or depressions that serve as sense organs, located in flexible appendages too small to be considered limbs.) But Joe wasn't trying to sound impressive; for him, tentacular fossae were simply tentacular fossae, and he paid the reader the compliment of calling them by their correct name. He was making an appeal to reason, the only sort of discourse that meant anything to him.

TWO years of the postdoc at LSU were followed by two years of teaching genetics and introductory biology there. In Louisiana, Joe led a simple, almost monastic existence, a choice dictated equally by rock-bottom pay and utter indifference to the comforts of middle-class life. When his father came to visit him in Baton Rouge, he found Joe living in an apartment that was almost completely bare. The apartment's sole claim to charm was a small balcony, which was reached by a spiral stairway. In the morning, father and son sat on the steps of the stairway to eat their breakfast, because there was no table. Ron said, "Today we'll go shopping, and I'll buy you a table."

Joe lost his temper. "Dad," he retorted, "I have everything I need in life. There's nothing I need that I don't already have." Ron Slowinski realized then how intently focused his son was on his vocation: "Joe had an uncanny intuition for career-building. He pared his life down to get rid of anything that didn't contribute to his career."

Almost monastic, but not quite: in Baton Rouge, Joe met Ashleigh Smythe, a young biology student from Virginia who would become his first significant girlfriend. She was bright, spirited, fit—and fine-looking, with a fresh, pretty face and a slim figure. They met in Baton Rouge in 1993 when Brian Crother invited Joe to speak at a seminar at Southeastern Louisiana University in Hammond, where Smythe was studying for her master's degree in biology. When a full professor asked a question after the lecture, Joe responded by saying, "Your question is poorly formulated." His answer made a poor impression on Smythe: She thought him a cocky young know-it-all.

Nonetheless, she said yes when he asked her out. Their first date was a Mardi Gras party given by Crother and his wife, Mary White. At the end of the evening, when Joe brought Ashleigh back to his apartment, he put on a spectacularly macho courtship display. Joe wrote in his journal: "I had locked my keys in the apartment and had to break down the door, completely destroying the door frame." He and Ashleigh began an intimate relationship, which continued off and on, mostly on, for five years.

In 1995, for the first time, Joe did not celebrate Christmas with his family; he and Ashleigh went on a camping trip in West Texas. Straying over to the Mexican side of the Rio Grande, the young lovers went prospecting for dinosaurs. "We were driving along the back roads in Mexico," Ashleigh recalled, "and Joe pointed at a hill and said, 'That one looks good.' We found a hadrosaur there. Joe was always good at finding stuff." They smuggled out two tote bags full of fossils, including a huge toe bone of the hadrosaur (also called the duck-billed dinosaur).

As conflicted as Joe was about living in Louisiana, stronghold of creationism, he loved the state's free-wheeling, high-living ethos. On weekends, Joe and Ashleigh would head for the bayous to go camping and herping. Joe would rattle down country roads in his Ford Ranger pickup, with the Ramones or ABBA blasting on the tape deck, drinking beer and tossing the cans out the window into the bed of his truck (but never littering the roadside). One of their favorite places to camp was on the grounds of Camp Shelby, an army base near Hattiesburg, Mississippi, which occupies part of De Soto National Forest. Once they were awakened before dawn by what they thought was an earthquake—then, moments later, a tank came rumbling over the berm, nearly flattening them.

Joe also loved the spicy cuisines of Louisiana. New Orleans, with its fabulous array of restaurants from funky to fine, was just over an hour away. He became a serious amateur chef and learned to cook authentic

Cajun and Creole dishes. One of the high points of Ron Slowinski's visit to Louisiana was a trip in the pickup deep into Cajun country. Joe took his father to a shrimp boil at a bar in the backwoods. They sat down and ordered beers; soon, a man came in from behind the counter with a big paper bag full of boiled crawfish and shrimp. Everyone dug in and shelled as many of the delectable crustaceans as they wanted. Cooking and eating good food, especially authentic ethnic cuisines, became Joe's lifelong passion.

At LSU, Joe had his first experience in the role of mentor. Frank Burbrink was a graduate student from Illinois who befriended Joe immediately after he met him at a departmental meeting in the herps lab. Burbrink recalled, "The other professors were all nicely dressed, but Joe was wearing a rumpled T-shirt with a cartoon of a grizzly bear chasing a man, and the caption ALASKAN DINNER TIME. I thought, This guy is pretty cool." Joe, for his part, would admit later that he was slightly intimidated by Burbrink, whose arms and chest are covered with ornate, intensely colorful tattoos.

After the meeting, Burbrink asked Joe where the best places were to go road-cruising for herps. Joe invited him to go snake hunting that night. When Burbrink showed up at Joe's apartment before they left for the woods, Joe said, "First, let's go to the Circle K." Burbrink asked why, and he said, "To buy some beer." That surprised the first-year student, but he soon became accustomed to this ritual. "Joe always drank beer, the cheap stuff, when he was out in the field," said Burbrink. "He was into this image of the redneck herper."

Joe took Burbrink to his favorite haunts. Once when they were collecting herps in De Soto National Forest, the state police stopped them. "There were empty beer cans in the back of the truck," said Burbrink. "The cops made us get out. One of them asked Joe, 'What are you boys doing out here?' Joe answered, 'We're biologists from LSU, sir. We're collecting snakes.' The cop said, 'I think you're out hunting cheap beaver.' " Joe looked him straight in the eye and said, "I have no idea what you're talking about, sir." He invited the skeptical policeman to pick up a specimen bag in the back of the truck, which held two eastern diamondback rattlesnakes caught earlier in the evening. The officer picked up the bag, eliciting a single, sinister-sounding rattle from one of its occupants. The policeman dropped the bag back into the truck and told the biologists to be on their way.

Despite Joe's ability to talk his way out of trouble, the Slowinski legend was taking on a dark tone. One day, Joe showed up at his favorite Mexican restaurant in Baton Rouge, a hole in the wall called Veracruz, with a pygmy rattlesnake in a plastic bottle. He had been drinking. He poured the snake out on the counter and told Ernesto, the owner of the restaurant, "I've got a rattler for you to cook. I'm going to eat it." Ernesto told Joe he was crazy and asked him to get the snake out of his restaurant. The little rattlesnake was about to escape, so Joe grabbed a ballpoint pen and pinned it down to the counter. The affronted snake turned around and bit his hand. Then Ernesto lost his patience and ordered Joe out of the restaurant.

Joe called Frank Burbrink from a pay phone in the parking lot and told him he had been bitten by a pygmy rattlesnake. Burbrink thought he was joking. Burbrink was about to leave for a concert in New Orleans: The Reverend Horton Heat, a popular rockabilly band, was playing at a club called the Howlin' Wolf. He told Joe, "You and Ashleigh come on and meet me at the Wolf." Joe picked up Ashleigh in his truck and drove the seventy miles from Baton Rouge to New Orleans. "By the time he got there," Burbrink said, "his hand looked like a blackened baseball mitt. He hadn't even gone to a doctor."

In Louisiana, Joe became habitually cavalier about handling venomous snakes. When he got home from a night of herping, sometimes he would just toss them in a bucket in the corner of his room, even though he slept on the floor in a sleeping bag every night. In Ashleigh's view, alcohol was impairing Joe's judgment. He wrecked her car one night while driving under the influence and had his license suspended. On another occasion he was arrested for public urination, on the lawn in front of his apartment building: He couldn't be bothered to wait till he got inside. When Ashleigh tried to talk to him about it, Joe denied there was a problem. "He thought I was being unreasonable," she said. "His attitude was, This is what herpetologists do. This is what macho field guys do."

While some herpetologists are quite conventional in their habits, most exhibit some degree of nonconformism, often verging on eccentricity. Chris Wemmer, a zoologist at the National Zoo (a part of the Smithsonian Institution), who began working in Burma in 1988, has published some informal essays about the personality of the zoo herpetologist, which apply equally to academic herpers. Wemmer believes that although they are often emotionally immature, herpetologists tend to be more committed and creative than other biologists. Those who specialize in venomous snakes are attracted to the field to some extent because of the kick they get from handling dangerous animals. Slightly (or sometimes excessively) embarrassed by this, they usually deny it and go to lengths to conceal their passion from outsiders.

Wemmer suggests that snake scientists tend to feel unappreciated and misunderstood, much like the animals they study. In a profile of the herpetologist, published in the *American Association of Zoological Parks and Aquariums Regional Proceedings,* Wemmer describes an adolescent boy (for until recently, the field was overwhelmingly male-dominated) consumed by a fascination with reptiles. He collects snakes and keeps them in his room, and uses them to stage scenes of predation for fun. The nascent herper also acquires the "symbols of pubescent machismo," such as oversize camouflage jackets and army boots.

"The syndrome becomes all-absorbing," writes Wemmer, "and the young herpetophile often develops a nerdishness which precludes membership in the school in-group. Some of these individuals fail to pass beyond this developmental phase: they become fixated on an interest that many of us were able to outgrow." Wemmer, who was trained as a mammalogist, also finds that herpetologists are generally of a higher intellectual caliber than biologists as a whole. "They understand evolutionary theory better, are more current in their taxonomic knowledge, and they cooperate with one another."

A colleague of Wemmer's at the National Zoo, herpetology curator James Murphy (Grace Wiley's biographer), also assembled a profile of the snake scientist, which was published in the *Herpetological Review*. Among the habits and traits he describes as being characteristic of herpetologists are indifference to wealth; keeping reptile collections at home from an early age; vigorous heterosexual behavior combined with an aversion to making romantic commitments; sloppy dress; a heavily carnivorous diet; a preference for wearing a moustache or beard; and a penchant for motorcycles. Murphy's profile fits Joe Slowinski perfectly, except that instead of a motorcycle he drove a pickup truck—another defiant emblem of old-school American manliness.

When asked to describe Joe, his friends always come up with the word "macho." His father said he was "masculine almost to a fault." Yet Joe was never the obsessive outsider Chris Wemmer describes. Although he was indifferent to being accepted by the in-group—in high school, or at any point in his life—he always had many friends. Moreover, his interest in the natural world was never confined exclusively to reptiles. When he was a child, he was a rock collector, which led him to the study of fossils. By the time he discovered his passion for snakes, he had already begun to think like a scientist.

In Louisiana, the gap between Joe's abilities and his professional status grew increasingly painful. By the end of his fourth year at LSU, it had been made clear to him that he would not be rehired as an instructor or offered a tenure-track position, so Brian Crother and Mary White contrived a one-year teaching position for him at Southeastern Louisiana. He was reluctant to take what had the appearance of being almost a charity post, but accepted it when, as he wrote in his journal, "it became apparent I would not otherwise have a job in the fall."

Joe's career was nowhere near the point he had envisioned it would be by now. A certain amount of floundering was normal for a new Ph.D., but he was no longer a newcomer. The stubborn fact was that he was six years out of graduate school, without a permanent job. A teaching job at Southeastern Louisiana University, while perfectly respectable, was approaching the lower end of the spectrum of the scientific establishment and would almost surely never lead to a permanent job. The post paid $27,000 a year, a pitifully small salary even in Hammond, Louisiana. Creditors were at his door: He was getting menacing letters from his car insurance company, a medical clinic, and a collection agency representing a record club. American Express was threatening to sue him.

The question "Now what?" tormented Joe more insistently than ever. He was doing everything right: He had published papers about snake systematics in good journals, some coauthored by Jay Savage; he was launching an online journal, one of the first, called *Contemporary Herpetology*; in 1995 he organized a symposium on snake phylogeny at the annual meeting of the American Society of Ichthyologists and Herpetologists, in Edmonton, Alberta. Yet while his growth as a theoretician was impressive, his fieldwork had dwindled to insignificance.

Chasing cottonmouths in the bayous was good fun, but Joe yearned to explore the tropical rain forests of Southeast Asia. He was already dreaming of leading an expedition to Burma, so long neglected by science, where he knew there would be spectacular biodiversity and new species waiting to be discovered. Joe knew an American ichthyologist named Carl Ferraris, who had discovered a new species of fish on sale at a public market there—the serendipitous stuff that science legends are made of. Joe knew that Burma would be equally rich with opportunities for a snake hunter like himself. It was the work he was born to do, but he couldn't carry out research like that without the support of an institution.

Time and again he came close to landing a good job, but somehow he always got the regretful rejection, even when his friends assured him he was far and away the most qualified candidate. Moreover, there was the depressing, very real possibility that he might never find his niche. Some qualified, talented scientists slip through the cracks, eliminated in the game of musical chairs that is modern academic life. Disappointed scientists constitute an informal league, like unpublished novelists and inventors who can't sell their gadgets: people who have complaints, often legitimate ones, that no one wants to hear. Some simply give up. David Good, the herpetologist who had supervised Joe's postdoc at LSU, moved to Canada to practice veterinary medicine after he failed to get tenure, despite being eminently qualified.

Joe confessed to his family and intimate friends that he had moments of black despair when he doubted if he would ever find a good post as a research scientist. Yet he carried himself as though he never lost hope. It was his destiny to make his name as a herpetologist. This had to be: Joe Slowinski refused to take no for an answer.

Kukri Snake

Oligodon arnensis

The kukri snake is a nonvenomous colubrid, commonly found in forests and near human habitations in South Asia, from Pakistan to Burma. It is an exceptionally muscular snake, with a prehensile tail and specially adapted ventral scales that enable it to climb rough vertical walls with ease. For defensive displays, it inflates its body to an enormous size by expanding its single lung backward into the chest cavity. The kukri snake is timid and only attacks if threatened, but it is often killed when it ventures into people's houses, on account of its superficial resemblance to the krait. The snake feeds chiefly on bird eggs and frogs, reptiles, and small mammals. Enlarged, backward-pointing teeth make escape extremely difficult for captured prey, which often die quickly from heavy bleeding.

When Joe interviewed for a job at the California Academy of Sciences, the committee asked him where he wanted to do fieldwork. He answered in a word: Burma. The logic was simple. Joe's research was now focused on the elapids of Asia; in order to learn more about them in the field, and particularly to find new species, he needed to get as deep as he could into wilderness terrain. And in 1997, few places in the world were more isolated and understudied than Burma.

An outcast from the world community ever since a military junta assumed dictatorial powers in 1988, postcolonial Burma was in an intellectual time warp. The last major research on the country's reptiles had been published in 1943, in the series *The Fauna of British India*, by Malcolm Smith, an amateur herpetologist who had been personal physician to King Chulalongkorn of Siam. By the end of the twentieth century, the jungle of Upper Burma was one of the last frontiers for field biology: If any place in the world was likely to harbor new species of cobras and kraits, Joe knew, this was it.

The committee in San Francisco liked what it heard. In the spring of 1997, just as his contract at Southeastern Louisiana University was ending, Cal Academy offered Joe the position of assistant curator of herpetology. His family and friends collectively sighed with relief: It was the perfect job for Joe. The Academy's huge collection made it an ideal platform for his wide-ranging work in snake systematics, and the position came with a robust budget for mounting field expeditions. The first person he called with the news, after his family, was Jay Savage. "Joe was really excited by the Cal Academy post," he said. "They told him they wanted him to do fieldwork. They said, 'Pick your spot.' "

Ashleigh, who was about to receive her master's degree in biology from Southeastern Louisiana University, decided to move with Joe to San Francisco. The two had broken up the year before, then gotten back together again after Joe moved to Hammond for the job at Southeastern Louisiana. He was ambivalent about her coming with him; he didn't come out and say so, but he seemed to want to start a new life in San Francisco. Frank Burbrink intervened and played the role of matchsaver, convincing Joe to stand by his woman. Before the spring term at

Hammond ended, Joe and Ashleigh flew out to San Francisco for a weekend and found a basement apartment in the Haight within walking distance of the museum.

Joe was joining a venerable tradition of scientific exploration at Cal Academy. Founded in 1853, it is the second oldest natural-history museum in the United States (after the Smithsonian, which preceded it by seven years). With total assets approaching $500 million and an annual operating budget of $20 million, it ranks fourth in the country after the Smithsonian, the American Museum of Natural History, and the Field Museum. The Academy was created in response to concerns about the impact of the California gold rush on the environment. From the beginning, it had strong holdings in herpetology: The first specimen in the museum's collection was the carapace of a Galapagos tortoise, catalogued in 1853.

In 1905, the Academy undertook its own collecting voyage to the Galapagos, where Charles Darwin honed his skills as a naturalist and began to spin his theories about natural selection. For evolutionary biologists, the Galapagos Islands are a shrine, the Stratford-on-Avon of natural history. The Academy's expedition there was led by a young curator named Joseph Slevin. While he was hard at work in the equatorial Pacific, gathering what would be one of the world's finest collections of reptiles from the remote archipelago, disaster struck at home: The great San Francisco earthquake and fire of 1906 destroyed the museum on Market Street, wasting the collection. Head curator John Van Denburgh arrived at the burning museum in time to save thirteen holotypes, the specimens used to describe and name new species. More than eight thousand carefully preserved specimens burned that day.

Slevin and Van Denburgh devoted the rest of their lives to rebuilding the herpetology collection, which was catalogued and stored in the California Academy of Sciences' imposing new Neoclassical home in Golden Gate Park. Van Denburgh amplified the museum's holdings by purchase; among others, he bought a big chunk of Edward Harrison Taylor's historic collection from the Philippines. After Van Denburgh's death by suicide in 1924, Slevin soldiered on until his retirement thirty-three years later, having led many expeditions throughout the world. Today, the Cal Academy herpetology collection comprises nearly three hundred thousand specimens, making it one of the largest in existence.

Joe Slowinski's research in Burma was a natural continuation of the work of Joseph Slevin and Edward Harrison Taylor. The Cal Academy herpetology department already had some seasoned explorers on the staff: Alan Leviton, who had been hired as a curator at Cal Academy in 1957 (the year that Slevin retired), had a wide-ranging career in the field, from the Arabian Desert to the mountains of Nepal to the Philippine Islands. Robert Drewes, the curator and chairman who hired Joe, had made twenty-four expeditions to Africa, where he discovered many new species. A former resident of Kenya, Drewes had led an expedition into the Impenetrable Forest of Uganda the year before Joe arrived.

In Joe's contract, Cal Academy undertook to pay for his move to San Francisco, but he had so few possessions that the movers in Hammond had a laugh when they arrived to pick them up. Joe drove solo out to California in his pickup, stopping in Taos to see his father. He took a long, challenging hike at Big Arsenic Springs on the upper Rio Grande, to see the petroglyphs incised in boulders by ancient hunters, and arrived in San Francisco a day late.

His first week on the job, Joe was ordered to represent the herpetology department at an event called a curators' forum. Anthropologist Christiaan Klieger, then the grants officer for the Academy, met him there. "It was basically a dog-and-pony show," said Klieger. "Everyone was there with their specimens, and the donors came to see what they were doing." All the other scientists participating wore business suits, but Joe was in his customary baggy T-shirt. Joe's presentation of himself as a hayseed genius was in part a self-romancing fiction—he was raised in a home that prized intellectual achievement—and in part an expression of his fear of competing in a sophisticated, competitive environment. "When I introduced myself to him," recalled Klieger, "he seemed intimidated, very shy. I thought, This poor guy needs some confidence-building. Little did I know, he had plenty of confidence in the field."

Joe's arrival was a big event at Cal Academy: New Ph.D. curators came along infrequently, and the herpetology department always possessed a certain mystique. After Leviton and Drewes had welcomed him, Joe met the rest of the staff. Jens Vindum, the bearded, apple-cheeked manager of the herps collection, would become Joe's indispensable partner in his research in Burma, and a close friend.

A few days after Joe started work, a tall, well-built Chinese man strode into the lab. It was Dong Lin, the museum's staff photographer. Abruptly, he said, "You're the new snake guy?"

"Yeah, I'm the snake guy," answered Joe.

Lin sized him up and said, "You're pretty tall. Do you play volleyball?"

"Sure, I play volleyball."

"Are you serious?"

"Yeah!" Joe grinned, amused by the direction the conversation was taking: no time wasted on frills, straight to the most important thing—sports.

"We play every Tuesday and Thursday."

"OK," said Joe, "I'm going to play with you guys."

He kept his word and showed up at the next game, at a volleyball court in the park. Lin was impressed by Joe's play. "He was pretty cool," Lin would later say, with the air of a man awarding the highest praise at his command. The game was a close one, and the closer it got, the more intensely Joe competed. "He wanted to win that game," Lin said. "In everything he did, Joe wanted to win. He wanted to show people he was rough and tough."

Joe became famous on the volleyball court for his devastating serve, a down-curving fastball that was very difficult to return; more often than not, it would score as an ace. Yet when it missed, it was wildly off the mark: The ball would go flying out of bounds or under the net to hit an unsuspecting opponent. Peter Fritsch, a botany curator at Cal Academy who would later collaborate with Joe on a major field project in China, was a regular on the volleyball court. He said, "That serve was a reflection of Joe's personality—he wanted to hit his goal directly, not in any roundabout or subtle way, and he was willing to take risks in the drive to succeed."

Dong Lin became Joe's best friend, following the pattern of Stan Rasmussen and Brian Crother: They worked together, played together, and drank together; on expeditions, they bunked together. When Joe finally decided to buy a suit, he asked Lin to help him pick it out. The photography department at Cal Academy was across the hall from herpetology. The two men often worked late, sharing a few beers from the photo lab's refrigerator before calling it a day; then they would move on to bars around Golden Gate Park for pool and deep talk. Their backgrounds couldn't have been more different: Joe, forever the white-bread Midwesterner, and Dong Lin, a child of the Cultural Revolution.

Five years older than Joe, Lin was born in Beijing and came of age amid the worst excesses of the Red Guards, the youth cadres that exercised dictatorial control over much of the People's Republic of China during the Cultural Revolution. When Mao closed the nation's schools in 1966, Dong Lin was just nine years old; to keep him out of mischief, his father gave him a camera. The boy was instantly in its thrall. Lacking access to a photo lab, young Lin had to find a way to develop and print his film. After years of saving his minuscule pocket money, he was able to scrape together twenty-five yuan, equivalent to a month's wages for most Chinese workers, which was enough to buy a lens he could use to build an enlarger.

When the universities eventually reopened, Lin enrolled at Beijing University. Although the curriculum offered nothing so bourgeois as a course in photography, there was a camera club, where Lin polished his technique. After he graduated from the university, he won a nationwide contest, and the prestigious Chinese Photographers Association, impressed by his work, invited him to join.

His first big story was as big as it gets: Lin was in Tiananmen Square in the spring of 1989, when the short-lived democracy movement occupied the great plaza. He got some appalling shots of the massacre on June 3 and 4—and more than sixty stitches for his injuries. Determined to get out of China, he obtained a passport, found a sponsor in the

United States, and sold everything he owned to come up with the money for the flight. Lin arrived in California, as he would later say, "with no money, no camera, and no English." He found a job washing dishes and saved his wages until he could afford to buy a Leica.

In 1992, Cal Academy hired Lin as a photo lab technician. After three years of cranking out black-and-white prints, he was promoted to the post of museum photographer. The nature of the job was changing: Lin's predecessors had been primarily responsible for documenting exhibitions and in-house research—stock photography suitable for brochures and publicity in local newspapers. But the Academy's new executive director, Patrick Kociolek, wanted Lin to accompany the museum's expeditions and create a visual record of the work of its scientists in the field.

Lin didn't fit any cultural stereotype of the Chinese that Joe knew: Far from presenting a cool, unflappable exterior, Lin was always in a flap, putting his emotions and brainstorms on display. He was at the mercy of his mercurial moods. Sometimes angry, always intense, he babbled a stream-of-consciousness monologue barbed with withering wit and punctuated by profanity. He and Joe soon became inseparable. Joe's previous sidekicks had been intellectual colleagues as well as playmates, but his friendship with Lin was founded on a boisterous brute empathy. Sometimes they roughhoused like adolescent jocks. As different as they were by culture, their characters were very similar: Both men were bright, ambitious, hardworking, and rebellious.

Joe left for his first trip to Burma in October, five months after he had started his new job. It was his first trip to Asia since the childhood year in Kyoto. Among his principal goals when he arrived at Cal Academy, perhaps the main one was to carry out a comprehensive investigation of the ecology of *Ophiophagus hannah*, the king cobra, the longest venomous snake on Earth; adults range from fifteen to nineteen feet in length. On his initial visit, Joe wanted to lay the groundwork for an intensive field research program, which would include implanting microtransmitters to plot the king cobra's range and taking blood from a large sample for population genetic analysis.

Just before he left, Joe learned that Wolfgang Wüster, the herpetologist he had met in Canterbury, was also planning a study of the species in southern Thailand. Joe sent him a quick e-mail, saying, "It sounds exactly like what I had planned, to the letter." He told Wüster that he was having doubts about continuing with the project: "Part of me does not want to reinvent the wheel. I guess I'll decide after the Burma trip."

If there were any objective measure of an animal's power to terrify, *Ophiophagus hannah* would vie with the tiger to head the list. Like cobras of the genus Naja, the so-called true cobras, when it is threatened the king cobra raises the upper third of its body from the ground and expands its hood. In this posture, the snake is five to six feet high, tall enough to stare into a man's eyes. Rather than hissing, the king cobra when aroused makes a loud, menacing moan, like the growl of a dog. To produce it, the snake fills its lungs with air, then constricts its body and forces the air through the glottis, the sliver of space between the vocal cords. The moan gains its deep, horripilating resonance when it passes through the snake's tracheal lung, an enlarged throat wrapped around the windpipe, which allows the king cobra to breathe while it's swallowing its prey.

The generic name Ophiophagus is Greek for "snake eater," which is a precise description: The snake's diet consists almost entirely of snakes, both venomous and nonvenomous, including even the robust Indian rock python. When there are no snakes available, the species occasionally eats monitor lizards (genus Varanus)—though not the largest of them, the ten-foot-long Komodo dragon, for the two species are not sympatric, meaning that their geographical territories don't overlap. (That didn't stop a low-budget production company in Hollywood from making a trashy science-fiction film in 2005 called *Komodo vs. Cobra*.)

Ophiophagus hannah has an extraordinarily wide geographic range—from the west coast of India to the Philippines, and from Indonesia north to the Himalayas. The first king cobra in Nepal known to science was captured in 1983 by a team of zoologists from Beijing. Yet despite its broad distribution, the species is endangered in many areas as a result of habitat destruction. Romulus Whitaker, an American-born herpetologist in India, has undertaken an ambitious program to study and help conserve the great snake. "I have been interested in snakes since I was a young boy of five or six," Whitaker said. "Growing up in the Western Ghats of India, I was initially fascinated with the king cobra on a very elemental level, because it's so huge and dangerous. As I grew up, I became interested in its unique ecology and behavior, such as the fact that it's the only nest-builder among the snakes."

Whitaker grew up in south India and studied in the United States, where he received a degree from the University of Wyoming and then apprenticed at the Miami Serpentarium with Bill Haast, whom he called his guru. He returned to India in 1967 and stayed on, eventually taking Indian citizenship. In 1976, Whitaker cofounded the Madras Crocodile Bank to support the conservation and study of India's three endangered crocodilians: the mugger, *Crocodylus paluster*; the gharial, *Gavialis gangeticus*; and the saltwater crocodile, *Crocodylus porosus*. Since then, the program has bred more than five thousand of the animals.

Yet Whitaker's professional passion has always been the king cobra. "It's an endangered species," he said, "which to me represents the rain forest, which is so dearly in need of protection." In more than thirty years of walking through the rain forest, Whitaker said, he has seen only six of the animals in the wild. In 2005, he established a research station in Agumbe Reserve Forest, northwest of Bangalore. There, Rom Whitaker and his team have found *Ophiophagus hannah* in relative abundance—not in its wonted wilderness hunting grounds, but near human settlements. "We find them in people's houses and gardens, in wells and pump houses, seeking shelter." Whitaker explained that wherever people live, there is food for rats; and wherever there are rats, there are rat-eating snakes—the king cobra's favorite food.

The species is revered as a forest god in many parts of India and Southeast Asia; some country folk believe that the serpent controls the rain and thunder. In one active cult at a sacred cave in Mount Popa,

Burma, a king cobra is brought to the village in ceremony in a basket. When the lid is opened, the snake rises and a priestess kisses it on the head three times, thereby guaranteeing both human and crop fertility. In Agumbe, Whitaker said, "the snake is worshiped more than anywhere else in India. Every household that can afford it has a stone sculpture of a five- or seven-headed king cobra. Thus fifty percent of our work is done for us—we don't have to teach people not to kill the animal, because they revere it."

Joe's first Burma trip was more like one of his adventures with Rachel or Brian Crother than the elaborate, ambitious expeditions he would later lead to the country. He arrived with no fixed itinerary. Like every first-time visitor to Yangon, Burma's capital (formerly called Rangoon), Joe went to see the Shwedagon soon after his arrival. It became a tradition for him: Every expedition to Burma began with a pilgrimage to the fabulous gold-plated pagoda.

More than the country's symbol, the Shwedagon is tantamount to the repository of the national soul. The immense monument spoke to Joe as it had to Western travelers for centuries. In 1586, an Englishman named Ralph Fitch wrote that the Shwedagon "is of a wonderful bignesse, and all gilded from the foot to the toppe. . . . It is the fairest place, as I suppose, that is in all the world." The bell-shaped stupa was constructed in the tenth century AD to house eight hairs from the head of Gautama Buddha. Rising three hundred and twenty feet from its base, on the top of a steep hill, the Shwedagon looms over the approaching visitor to a height equivalent to that of the Pyramid of Cheops. For centuries, the devout of Burma have continuously replated the surface of the stupa with gold, which is now estimated to weigh well over one hundred thousand pounds. The ornamental crown, called the *hti*, set with thousands of diamonds, rubies, sapphires, and topazes, may be seen glittering from miles away.

In Burma, Joe quickly learned to curb his habitual outspokenness and maintain a low profile. Throughout Yangon, menacing scarlet signs on the roadsides urged citizens to inform on neighbors and friends who criticized the government. Most people were afraid to be seen talking with foreigners, and those who were bold enough to do so never discussed politics. Yet Joe felt an instant sympathy with the warm-hearted, intellectually curious Burmese people and made friends everywhere he went.

A few days after his arrival in Yangon, he struck up a conversation with a Burmese businessman at his hotel. The man told Joe that he had just returned from the south, where a ferocious, growling king cobra had pursued him and his family, forcing them to spend a terrifying night locked in their car to escape the strikes of the aggressive predator. Joe got excited and explained what he was doing. The man directed him to a village in the Irrawaddy Delta called Mwe Hauk. The name meant Cobra Village, he said—and the place lived up to it.

Joe hired a car and headed straight for Mwe Hauk. It was the middle of a monsoonal downpour, much too heavy a rain to go looking for herps, so he spent three days holed up in a leaky bamboo house in the village, eating wretched food (including house-cat curry). When the rain finally let up, Joe grabbed his snake stick and collecting bags, intending to head for the jungle. He never made it. Word had gotten around about what he was doing, and one villager after another came to him with specimens for sale. As Joe told the story in an article in the *San Francisco Examiner Magazine*, "All hell breaks loose. People bearing snakes, lizards, and frogs arrive from everywhere. I am stunned." A child not more than seven brought him a monocled cobra; an old man came in with a juvenile *Ophiophagus hannah*. By day's end, Joe had bagged more than sixty snake specimens, along with scores of lizards and frogs.

"I am elated at the haul," he wrote, "but my back aches from stooping over specimens, and I am definitely looking forward to kicking back with a beer or two. But before I can, I hear a loud commotion—chickens squawking. I check their bamboo hutch; a kukri snake has crawled in to eat some eggs. I bag the snake, and think what a great country Myanmar is."

Golden Tree Snake

Chrysopelea ornata

The largest of the so-called flying snakes, *Chrysopelea ornata,* common in lowland rain forests in India and throughout Southeast Asia, cannot actually fly; rather it leaps from high places and glides for distances of up to a hundred yards. To accomplish this, the snake flattens its entire body, which is three to four feet in length, forming a U-shaped concavity. The outer edges of the ventral scales are hinged, so they can be raised to enhance the parachute effect. The snake launches itself from a coiled position, snapping its body forward through the air in a controlled glide. It undulates through the air as if swimming, holding its tail rigidly upward and twisting it from side to side for stability. The golden tree snake (which is actually bright green in color) usually "flies" from tree to tree, but sometimes from tree to ground; it can climb quickly to a new launch point, aided by ridged scales on its belly. *Chrysopelea ornata* is mildly venomous, an adaptation for hunting small prey such as lizards, frogs, birds, and bats.

Joe came back from Burma besotted with the place. In his article for the *San Francisco Examiner Magazine*, he wrote:

> Snakes are not the only attraction in Myanmar. The country is visually stunning. In this devoutly Buddhist nation, pagodas and temples, many of them hundreds of years old, dot the landscape. The enormous Ayeyarwady [Irrawaddy] River cuts through the middle of the country, unconstrained by dams and levees. And jungle-clad mountains form a continuous, horseshoe-shaped barrier along the borders on three sides, walling the country off from its neighbors. Furthermore, the people are unbelievably friendly, and crime is virtually unknown.

There's more than a touch of the neophyte's naïveté here: Just about every place in Southeast Asia is well-supplied with ancient temples and smiling faces, and Burma harbors many criminal gangs (including, some would say, its government). Like so many naturalists before him, Joe was spellbound. George Zug said, "Joe was smitten by the Burma Bug. He had a romantic view of Asia, like Kipling." Joe had a lot to learn, but he was a dedicated student. Over the course of four years, he would visit the country eleven times.

His most important accomplishment on that first trip was making a new friend in U Uga, the affable director of the Nature and Wildlife Conservation Division of the Forest Ministry. Joe called on him in Yangon to discuss the possibility of setting up a field study of the king cobra, though he still wasn't sure if he wanted to do that. (He never did.) U Uga made him a more grandiose counterproposal: He invited Joe to undertake a comprehensive survey of the country's herpetofauna, a scientific census of all its reptiles and amphibians. As an inducement, U Uga dangled the alluring prospect of an expedition into uppermost Burma, to the virtually unexplored region around Hkakabo Razi, which at 19,250 feet is Southeast Asia's highest mountain. The Forest Ministry, with the aid of the Wildlife Conservation Society, had recently established a nature preserve there. Joe e-mailed this report to Brian Crother after his return:

Dude,

Burma was great. I collected about 100 snakes, including cobras, kraits, Trimeresurus [a genus of Asian pit vipers], *green vine snakes, etc. I am hoping to go back in 2 months. It looks like the Burmese government is giving me permission to go to this newly formed national park in the north. It's all virgin montane forest—never been collected, not even by the British 50 years ago.*

It was exactly the sort of ambitious research project Joe was looking for, yet, surprisingly, he didn't jump at the idea. He wanted to get back to Mandalay as soon as possible. He was on a quest.

In the same e-mail exchange with Joe in which he revealed his own study of the king cobra, Wolfgang Wüster told him about the probable existence of an unidentified snake species in the Mandalay region—a spitting cobra. In the 1980s, when Joe began his graduate studies at the University of Miami, science recognized but one species of cobra in Asia (apart from the king cobra), *Naja naja*. In twenty years, it would multiply, Hydra-like, into eleven species.

Wüster had been leading the effort to rationalize the taxonomic muddle of Naja. He had already predicted the existence of the Burmese spitting cobra five years before, in a paper for the journal *Herpetologica*, which was coauthored by Roger Thorpe of the University of Aberdeen. This study of the fangs of Asiatic cobra species and their adaptation to spitting behavior found that, contrary to expectation, many spitters have short fangs; the key adaptation is a narrowing in the orifice of the venom discharge duct, which has the same power-boosting effect as pinching a thumb over the end of a garden hose.

For Joe, perhaps the most interesting aspect of the paper was its résumé of field observations of spitting behavior. Reports from Indonesia, Thailand, Malaysia, and the Philippines were fairly recent and appeared to be sound; there had even been some reports from the field in Cambodia by a French herpetologist as recently as 1972. Yet the solitary field report from Burma dated to 1900—a description of a spitting cobra in Mandalay, written by a lieutenant in the British Army named Michael

Goring-Jones. His paper was entitled "Can a Cobra Eject Its Venom?" It was a lame question even for a Victorian amateur; as early as 1827, a German herpetologist had been so impressed by the spitting abilities of the Javanese cobra that he named it *Naja sputatrix*. Yet just two species of cobra were known to exist in Burma: the king cobra, definitely not a spitter, and the monocled cobra, *Naja kaouthia* (officially still classed as a subspecies of *Naja naja*, like all Asiatic true cobras). Wüster had examined many specimens of *Naja kaouthia* in Thailand but had never seen one exhibit spitting behavior. There were a few stray reports of spitting monocled cobras, but they were just as uncertain as the description by Goring-Jones. He was a soldier, not a naturalist; perhaps the cobra he saw had simply hissed and sprayed a cloud of saliva, as many snakes do.

Or, intriguingly, Goring-Jones might have chanced upon an altogether different species, as yet unidentified by science. At the Stockholm Natural History Museum, Wüster had seen six cobra specimens from Burma with fangs well adapted for spitting, which he couldn't assign to any of the known Asiatic cobras. They had been collected by an ichthyologist in 1935, who labeled them "Rangoon or Mandalay"—a worthless locator. Wüster and Thorpe concluded, "It appears that a species of cobra with spitting fangs occurs sympatrically with *Naja kaouthia* in parts of Burma."

For Joe Slowinski, Wolfgang Wüster's offhand tip and article in *Herpetologica* possessed the allure of a map for buried treasure. The existence of a new species of cobra had been convincingly deduced from some poorly labeled, sixty-year-old specimens in Sweden and a century-old eyewitness report by a soldier-dilettante. Joe wanted to be the one to discover this spitting cobra of Rangoon or Mandalay.

He was back in Burma four months later, in February 1998. Jens Vindum, Cal Academy's hardworking, easygoing collections manager, came with him. Joe tried to recruit Dong Lin to join the group, but he was already booked to accompany an entomology expedition in Madagascar. Joe wheedled, "Why do you want to go to Madagascar to take pictures of insects? That's boring! Come to Burma and shoot snakes. That's man's work." He taunted: "Ask your girlfriend if you can come to Burma."

Also joining the second expedition were two young female biologists: Curatorial assistant Carol Spencer flew to Burma with Joe, and a graduate student from Harvard named Heidi Robeck would meet them there. A quiet revolution in herpetology had begun by the end of the twentieth century: It was among the last scientific fields to include women in significant numbers. Kraig Adler's *Contributions to the History of Herpetology* contains 151 mini-biographies of great thinkers in the field, from Conrad Gessner (1516–1565), the author of *Serpentium Natura*, the first work to treat reptiles as a distinct class of animals, to Avelino Barrio (1920–1979), an Argentinean venom expert. Of them, only five are women, all twentieth-century frog specialists. Carol Spencer, whose studies focused on rattlesnakes, said, "The attitude was always, 'Girls don't study snakes, girls study frogs.' When I started working on rattlesnakes, there were four women doing it."

There are two notably courageous female herpetologists in Adler's book. In 1918, at the age of thirty-five, Alice Boring left a good job at the University of Maine to teach biology in Peking. She continued working in China until 1950, with a few years off for teaching posts at home and a stint in a Japanese internment camp. Brazilian herpetologist Bertha Lutz (1894–1976) studied the Hylidae, a huge family of tree frogs. She was also a lawyer and one of the founders of the women's rights movement in Brazil. She was a member of the committee that drafted the country's new constitution, which gave the ballot to Brazilian women for the first time in 1933.

While they waited for their permits to come through, Joe took Vindum and Spencer to see the Shwedagon and the capital's other sights. As soon as they got their papers, they flew to Mandalay. There, on the grounds of their hotel, Joe caught a flying snake, *Chrysopelea ornata*, by snatching it from the bough of a tree.

The herpetologists began collecting in Pakokku, a village renowned for its clan of snake catchers, men who capture and remove venomous snakes from villages and fields, where they pose a constant threat to human life. The snake catchers' lean, sun-blackened bodies are covered with magic tattoos, tinted with ink infused with snake venom. They believe that the tattoos make them invulnerable to the venom's toxic effects.

There was no way of knowing what habitat the conjectured spitting cobra might favor; anyway, Joe and his team were there to collect whatever they could find. The terrain around Pakokku appeared to be utterly un-snaky: bare, packed dirt, disused paddy fields with no trees to shade them. Yet the Burmese snake hunters, wearing only *longyi* (traditional Burmese sarongs) and plastic sandals, were soon plucking large Russell's vipers from the brick-hard earth with improvised bamboo tongs. The venom of *Vipera russelli* is more toxic than that of the cobra (yet less toxic than that of the many-banded krait); very few victims survive the snake's bite. It's a horribly painful death, typically resulting in massive bleeding from the anus and in the brain, and renal failure. Those who do survive are wrecks: The venom plays havoc with the pituitary gland and reverses adolescence, turning the victim back into a physical eight-year-old. Secondary sexual characteristics recede, leaving the sufferer smooth of body hair, impotent, and sterile.

When the group moved on to Mwe Hauk, where Joe had captured the kukri snake on his previous visit, they stayed at the house of the chief. Soon after the group arrived, the chief challenged Joe to an arm-wrestling match; such mano a mano trials of machismo would become a common ritual as Joe traveled through rural Burma. Just as on his first visit to the village, specimens came raining in from the local folk. There were all sorts of intriguing items in the initial haul: a venomous sea snake, some rare water snakes, the deadly many-banded krait, and the lovely green *Ahaetulla prasina*, a gracile vine snake that curls itself into elegant arabesques.

Very few people in Mwe Hauk had ever seen Westerners, so golden-haired Joe, impressively bearded Jens Vindum, and a pair of attractive young American women drew crowds everywhere they went. As the scientists sat in camp laboriously preparing and photographing their specimens, villagers pressed up near them, rapt with attention. Joe and the others were at a loss to understand how such boring, repetitive work could be so compelling, but the whole time they were there, a pack of locals stood around them, observing their every move. It got on their nerves.

The Americans then returned to Mandalay for a training session with the Forest Ministry staff. On the first day, U Uga had too much to drink before the opening ceremony and got so intoxicated that he let his *longyi* fall down while he was introducing the American guests. Joe met Htun Win and Thin Thin, the biologists who would become the nucleus of his in-country field team.

Shortly before the end of his stay, Joe got lucky—not in the field, but at an illegal snake-exporting business. Commerce in snakes had been outlawed in Burma, but the trade still prospered in the dry zone around Mandalay. The snakes were captured by amateur collectors, mostly subsistence farmers, who sold them to wholesale businesses that trucked them to China for sale at a premium price as exotic food or for "medicinal purposes" (which usually meant as an aphrodisiac for men). In 1999, Joe estimated that 1.6 million pounds of snakes were being exported illegally from Burma every year. He often stopped to have a look at the dealers' stock, to see if they had anything interesting.

Joe later described what he found in an article for *California Wild*, published in 2000: "I get excited when a dark brown cobra, which we are told was caught locally, is brought out." Joe saw at once that it was neither a monocled cobra nor a king. The former have the distinctive hood marking that gives it its name, while this specimen was patternless; and it was obviously not a king cobra. Then, while Joe was examining the snake, the confirmation came: No one was ever gladder to be spit at by a cobra than Joe Slowinski was that day, in the godown of an illicit snake smuggler in Mandalay. It was the new species.

The gap in scientific knowledge about Burma yawned back to the earliest exploration of Asia. Burma had been drawn into the British sphere of influence by the East India Company centuries before it became a part of the Empire. However, Lower Burma, the southern part of the country with its capital in Rangoon, wasn't brought under British control until 1852; Upper Burma, ruled from the ancient capital of Mandalay, was finally annexed in 1885. Burma at that time was considered a province of India, and looked down on as backward by the British intelligentsia. There were occasional visits by intrepid amateurs, and university-trained military officers like Michael Goring-Jones collected wildlife sporadically, but imperial biology mostly neglected Burma.

One of the most productive scientific investigators in British Burma was Italian. Leonardo Fea (1852–1903), an explorer from the Natural History Museum in Genoa, spent four years there beginning in 1885, creating a large collection of snakes, birds, and insects. One of the rarest snakes in Asia is named after him: the mildly venomous Fea's viper, *Azemiops feae*. Identified from only a few dozen specimens, Fea's viper is a monotype—the only known species of a subfamily of the Viperidae that was created especially for it.

In the twentieth century, botanists were more active than zoologists. Frank Kingdon-Ward towers over them all in terms of sheer productivity, with a career spanning forty-eight years: In ten epic journeys to Burma from 1914 to 1956, he discovered dozens of new plant species and collected hundreds of varieties of begonias, poppies, rhododendrons, and other showy flowering plants, which became staples of English gardens. His carefully observed, dryly witty accounts of his journeys, with titles like *Burma's Icy Mountains* and *In Farthest Burma*, were popular reading for British Sunday gardeners. One of his competitors, Reginald Farrer, discovered dozens of species of rhododendron in Burma; in 1920, at the age of forty, he died of diphtheria on an expedition near Putao.

In addition to the extreme health risks, travel in Burma was enormously complicated by the country's nearly chronic state of conflict. A pair of British botanists, Ronald Kaulbeck and John Hanbury-Tracy, set out on a remarkable expedition in 1935 to seek the source of the Salween, Burma's longest river. They walked nearly three thousand miles in twenty-two months, following the river's course through the steep gorges that slice through the Shan Plateau, but had to turn back before they reached their goal when they encountered a local political dispute that had turned violent.

British control over much of Burma was always wobbly. In the late 1920s, a Buddhist monk named Saya San led an uprising against the

British, demanding restoration of the Burmese monarchy. Saya San's followers went into battle armed with spears, daggers, and magic tattoos, which they believed made them invulnerable to the bullets of British machine guns. By the time Saya San was captured and hanged, more than ten thousand of his followers had been mowed down. The northern border territories, including modern Kachin State, where Joe led his expedition in 2001, were shaded pink on the map, but they were never effectively ruled by the British (or anybody else) for long.

When independence came, in 1948, the new nation was called the Union of Burma to acknowledge its loosely confederated constitution, but its unruly ethnic factions never lived peacefully with one another for any length of time. The country's hopes for stability were shattered from the start by the assassination of its charismatic war leader, General Aung San, who had headed the transition government. The ineffectual prime minister, with the tiny, palindromic name of U Nu, turned power over to his top general, the erratic Ne Win, formerly Aung San's second in command. Under Ne Win's despotic rule, Burma embarked on a disastrous course of extreme isolationism and a planned economy called the Burmese Road to Socialism.

Ne Win's Burmese state was characterized by harsh suppression of dissent coupled with larcenous corruption. In 1988, a series of mass protests by workers and students in Rangoon brought the country to the edge of revolution. The army brutally broke up the protests, killing thousands of people, and seized control of the government. This coup, which most historians believe was organized by Ne Win, installed a junta called the State Law and Order Restoration Council, usually known by the risibly sinister, Bond-ish acronym SLORC. The generals began better than anyone expected, by implementing economic reforms and calling for a free election in 1990; but when the National League of Democracy, led by Aung San Suu Kyi, Aung San's daughter, won in a landslide at the poll, SLORC annulled the election and put the democracy movement's icon under house arrest.

Burma became a pariah in the international community. Foreign scholars faced formidable barriers to working there—not only from xenophobic SLORC but also, in the case of most Western scientists, from their own countries, which imposed strict sanctions against dealing with the regime. Burmese scientists, too, faced obstacles from their military overlords. In the SLORC era, the nation's universities were under constant surveillance and suffocating ideological control by the generals. In 1996, they shut the universities down completely—the first of several such interruptions in the nation's intellectual life.

Nonetheless, venturesome biologists found a way around these obstacles. In 1988, a delegation from the National Zoo, led by Chris Wemmer, visited Rangoon to set up a scientific partnership, but the national strike and the subsequent rise of SLORC scotched their plans. Four years later, thanks to the vision and energy of U Uga, the door opened again. The Smithsonian was first in: Chris Wemmer set up a wide-ranging collaboration, beginning with a field research project to study the ecology of the endangered thamin, or Eld's deer (*Cervus thamin eldi*).

In 1996, Alan Rabinowitz, the director of exploration for the Wildlife Conservation Society, the research institution of the Bronx Zoo, helped U Uga and the Forest Ministry set up Hkakabo Razi National Park. Rabinowitz, a mammalogist, wasn't allowed to do fieldwork there himself until the spring of 1997 (a few months before Joe Slowinski's first trip), when he climbed into the foothills of the Himalayas with a small group of Burmese forest rangers. Even on this relatively quick trek, he discovered a new species of leaf deer.

Joe's timing was flawless: He was making his first bold move just as the scientific territory in Burma was being staked out. Soon, however, he would discover how byzantine and acrimonious the politics of Big Science can be: Alan Rabinowitz would prove to be a formidable adversary to Joe in his research in Burma.

Joe was elated to return from his second trip to Burma with a new species of cobra in his luggage, but none of the snakes he had bought at the illegal exporter's depot could serve as the holotype, or type specimen, of the snake he and Wolfgang Wüster would describe in a scholarly paper identifying the new species. The main objection was that there was no record of precisely where those snakes were found; with the easy availability of cheap devices to access the Global Positioning System (GPS), modern science demanded precise coordinates. It was also a matter of pride: Joe wanted to catch the holotype of the Burmese spitting cobra with his own hands.

Almost from the day he landed at Cal Academy, Joe's professional focus was razor-sharp. The sense of drift during his final years in Louisiana was replaced by intense concentration: After he returned from his second Burma trip, Joe worked with lean, stripped-down efficiency. In addition to his duties at the Academy, he was negotiating a major, long-term scientific partnership with the Burmese and seeking the funding for it. At the same time, he took charge of every aspect of *Contemporary Herpetology*, the online journal he had started when he was in Hammond, recruiting distinguished contributors who would get the journal noticed.

His growing obsession with his work was putting a strain on his relationship with Ashleigh. She had a brutal reverse commute from their apartment in San Francisco to a biotechnology lab in Sunnyvale, more than forty miles away, where she started work at six a.m. Joe was a night owl and would often play loud music as he worked, while she was trying to sleep. "He wasn't so considerate," she would recall years later, without rancor. "It was hard for Joe to see things from another person's point of view." There may have been a limit to how far the relationship could develop: No one's opinion carried more weight with Joe than his mother's, and Martha had never liked Ashleigh. Martha Crow had left Kansas City to return to New York, where she attended graduate school and eventually found a job as an editor at *Food & Wine* magazine. Christmas 1997 was tense; Ashleigh inadvertently provoked an argument with Joe when she offered Martha some unsolicited advice about salad-making.

Following what would soon become a pattern, Joe began planning his next trip to Burma almost as soon as he had unpacked from the last. This time, the turnaround was just three months. In addition to renewing the search for the Burmese spitting cobra, he had promised U Uga that he would give a series of training sessions for park rangers and Burmese biologists at Alaungdaw Kathapa National Park, near Mandalay. Jens Vindum and Heidi Robeck were both returning to participate in the training program, but Joe, impatient, went ahead on his own.

He called on no one in Yangon and went straight to Mandalay. Accompanied only by a driver and an interpreter, he first headed west, to Monywa, on the Chindwin River. Monywa, in the country's central dry zone, is one of the hottest places in Burma, and May is the hottest month: Temperatures top a hundred degrees Fahrenheit every day. The terrain was scrubby savanna, parched dry. Joe quickly saw that it was useless for herping; in such extreme heat the snakes would be sleeping in empty rodent holes. He headed north in search of more promising territory. He stopped to ask about cobras at every village he passed. Everyone he met seemed to have a story to tell about a relative or friend who had died of a snakebite. At one village they told Joe that a cobra had been killed just the night before, but no one was sure what had become of the carcass. Come back later, they said.

Joe returned that night, when the land had cooled to a point that snakes might be active. He was the first foreigner to come through in a long time, so there was the usual crush of curious onlookers. A man had found the cobra that had been killed and brought it to Joe. It was the spitter all right, which was excellent news. Unfortunately, mangled by machete blows and already decomposing, it was useless as a specimen.

Joe started hunting for himself, every nerve focused on finding the spitting cobra. Doubt was beginning to give way to dejection when he heard the voice of a woman somewhere in the village, screaming *"Mwe! Mwe!"* Joe raced through the maze of dusty lanes and bamboo thatch-roofed houses toward the source of the cries, a hundred Burmese villagers running right behind him. A frightened family had cornered a small spitting cobra in their hut. "After some awkward thrusts," Joe wrote in his article in *California Wild*, "I snag it with my snake tongs and secure it in a cloth bag. Finally! Elation!"

The villagers, who understood how important the snake was to their foreign guest, laughed and crowded around Joe to congratulate him; he shook a hundred hands. Joe was happy but far from satisfied. After noting down the GPS coordinates for his specimen, he asked his guides to take him out into the fields to look for more cobras. The normally obliging Burmese disregarded his request and triumphantly led him from house to house to drink toddy, a potent wine made from the sap of a domesticated palm. The villagers had made Joe's discovery the occasion for a drinking party and had no intention of taking him on any more dusty snake hunts. After shaking the hundred hands a second time, Joe climbed into his car to continue on his way. Two hundred feet down the road, he found a Russell's viper stretched out on the pavement, which he promptly snagged and bagged.

When he returned to Yangon for the training sessions, Joe met with George Zug, who was also leading a training program under the auspices of the Forest Ministry. Zug was staying at Joe's hotel, the Highland Lodge, a simple, cheerful place located far from filthy central Yangon. The hotel's owner, a portly, hospitable man named Tin Htut, who was getting ready to retire to Australia, looked upon Joe as his adopted American son. At one point, Joe kept a twenty-foot-long Burmese python in the lodge's pleasant little garden.

By now, Joe had resolved to set up a permanent presence in Burma for the California Academy of Sciences, just as Chris Wemmer had done for the Smithsonian. Joe had fallen hard for Burma, and he now regarded U Uga's proposed nationwide herpetofauna survey not only as a rich scientific opportunity but also as a way of traveling the country. Most macho field scientists disdain any interest in crumbling temples, but Joe, perhaps influenced by his father's study of American Indian civilization, felt an affinity with the country's vibrant culture that rivaled his passion for its herpetofauna. The drafty cathedrals and hushed museums of Europe had left him cold; he was on his way to becoming an old Asia hand.

Joe's professional rise had been marvelously swift: Less than a year earlier, he had been an untenured lecturer at Southeastern Louisiana University, and now he was launching a major international scientific collaboration. The project needed some instant gravitas, and George Zug was one of the undisputed heavyweights in the field of herpetology. Zug's textbook had quickly become the standard work for undergraduates, putting his mark on the emerging generation of researchers. He had also received the ultimate honor for a zoologist, having a species named after him: *Pseudemys gorzugi*, a cooter turtle in the Rio Grande Valley.

In their meeting at the Highland Lodge, Joe urged Zug to join forces with him. Zug resisted at first, but eventually came around: Joe possessed great powers of persuasion, and he never gave up. The two of them went to U Uga to negotiate a deal. They agreed on a cooperative long-term agreement between Cal Academy, the Smithsonian, and the Nature and Wildlife Conservation Division, with three main goals: to train Burmese biologists in herpetology, to survey Burma's reptile and amphibian life, and to use the data gathered by the survey to promote conservation of the animals. The way U Uga talked, there would be no barriers to Joe exploring anywhere in the country. He repeated his invitation for Joe to lead an expedition to the Hkakabo Razi region.

Joe was forming a close circle of Burmese biologists who would become the core of his field team in the years to come. The most talented of the group were Htun Win and Thin Thin, park rangers who had trained with both Chris Wemmer and George Zug. They were a couple; they had met and fallen in love in Wemmer's course in 1996, and later married. Joe recognized a kindred spirit in Htun Win: If anything, the young Burmese herper was even better at chasing down and catching snakes than he himself was. Win was remarkably dedicated, often staying up until midnight to study and work on his specimens, then rising at dawn to continue. And Thin Thin, Wemmer said, had "supernatural vision when it came to seeing wildlife." Even from a great distance, she could spot cryptic species so well camouflaged to disappear in the litter of the jungle floor that trained observers might have stepped on them.

On his way home to San Francisco, Joe stopped over in Japan for a sentimental journey to Kyoto, armed with detailed instructions from his father about how to find the house on the Takano River where he had lived when he was six. He wrote in his journal, "I set off to find our old home. Apparently, it doesn't exist anymore. I walk about town myself, marveling at old Shinto shrines and food markets. I walk to the old imperial palace—I remember walking here as a child and being disappointed by the sterility of the palace moat."

Joe's third expedition to Burma, brief as it was, discovered several new species in addition to the holotype of the Burmese spitting cobra. In Alaungdaw Kathapa National Park, he collected the type specimen of a new species of wolf snake, which, in a mild fit of sycophancy, he named after U Khin Maung Zaw, U Uga's successor as head of the wildlife division: *Lycodon zawi*. The team also found a new species of bent-toed gecko that scampered over the rocky banks of the park's rivers. Years later, it would be named for Joe Slowinski.

In San Francisco, Joe's top priority was to get DNA sequence data from the tissue of the Burmese spitting cobra so he and Wolfgang Wüster could compare it with other Asiatic cobras. The results confirmed their prediction: It was a new species. The snake he captured in the hut that night in Monywa would be the holotype. As they collaborated on their paper, later published in *Herpetologica*, Joe and Wüster agreed about every major issue—until it came to the naming of the species.

Joe insisted that they name it *Naja myanmarensis*. Like every foreign intellectual who worked in Burma, Joe had to come to some sort of ethical reckoning about doing business with the nation's cruel, corrupt regime. He wrote a clear account of his conclusions in an e-mail to Wüster:

> *One might ask, why go through all the trouble to cooperate with an oppressive military government? But the truth is there are some government*

biologists within the Forest Ministry that truly care about conservation and biodiversity, and that are trying to slow the high rate of deforestation. I feel that my efforts can strengthen the resolve of these individuals and help to spread the enthusiasm for ecosystem conservation through the ranks of Forestry. With any luck, my work will result in some good. Thus, I kowtow to the junta.

Joe thought it would encourage his friends at the Forest Ministry—and incidentally bolster his own standing there—if he named the new cobra species after the country. Wüster replied that he was dubious about "*myanmarensis*," fearful that it might suggest that the authors supported the regime that had recently adopted Myanmar as the country's name. Nonetheless, he agreed to let Joe have his way. At the end of his message, as an afterthought, Wüster tossed off "*mandalayensis*" as an alternative. Joe liked it, and that's what the snake was called. It was the first new species of cobra to be described since 1922.

Finding new species wasn't necessarily the principal aim of Joe's expeditions to Burma, but it was their most glamorous activity. Expanding the known range of established species was also an important goal. In a larger sense, helping to create a self-sustaining field biology program in the country, led by talented young naturalists such as Htun Win and Thin Thin, would ultimately make a greater contribution than finding any particular lizard or snake. Yet at a gut level, nothing could match the thrill of identifying a distinct, clearly defined member of the animal kingdom that had existed for millions of years, long before *Homo sapiens* walked among them. It was tagging a tiny piece of Creation as your own.

In *California Wild*, Joe wrote, "For a herpetologist, finding a new species is always exciting; for me, finding a new cobra species is the ultimate discovery."

Burmese Spitting Cobra

Naja mandalayensis

Closely related to the monocled cobra, with which it is sympatric, the Burmese spitting cobra is found in the arid zone around the city of Mandalay. Adults grow from three to just under five feet long and are medium to dark brown in color. Like most species that exhibit spitting behavior, *Naja mandalayensis* aims a stream of venom at the eyes of a predator to stun and disable it. Since the species was formally identified, in 2000, there have been no studies of the ecology of the Burmese spitting cobra. The species is not endangered but is under increasing pressure on account of being intensively hunted for the traditional Chinese medicine market.

Joe Slowinski was ambivalent about fame and fortune. He was certainly ambitious in his own field and possessed a healthy ego: A part of him thirsted for recognition. Yet since the Victorian era, Western science has been dominated by an ethos of modesty. Scientists live a life of the mind, so goes the paradigm, too deeply immersed in unraveling the mysteries of the universe to chase after fame. Glory may be thrust upon you, if you are truly great, but the preferred response to lionization is to slouch into your scholarly den and be reclusive, coolly enigmatic to any civilians who seek you out.

The archetype was Darwin, in his magnificent seclusion in the Kentish countryside, equally indifferent to the lampoons of his theory of natural selection by his detractors and his elevation to the status of intellectual demigod among his colleagues. In the electronic age, prominent scientists might give interviews to journalists and even appear on television, as Stephen Jay Gould did on many occasions on public-affairs programs, to defend evolutionary biology against the assaults of creationists. But for a scientist to seek renown actively is a sure way to be ostracized by his peers. A "real" scientist will go to any length to avoid being called a popularizer.

Joe's eccentricities—the redneck affectations, the penchant for punk and disco music, the futon-on-the-floor asceticism—fitted easily within the tolerant ethos of contemporary academic scientists, particularly herpetologists, who gave up safari suits and pith helmets long ago. At the same time, subsumed within Joe's defiant individualism was a deep reverence for the noble tradition of the natural historians who had preceded him. From an early age, he had virtually worshiped Darwin. Joe's cultivated nonconformism was a veneer that covered both a high seriousness of purpose and a powerful need to be accepted by his profession. Yet when Brady Barr, Joe's old friend from graduate school, began to make his name as a rugged on-screen herpetologist at National Geographic Television, Joe wasn't shy about hitching his wagon to the rising star.

Soon after he was hired by National Geographic, Barr called Joe to brainstorm ideas for television programs based on Joe's research. One morning in January 1999, Barr took what they had come up with to executives at the *National Geographic Explorer* series and fired their interest. That afternoon, Barr sent a message to Joe, asking him to pull together some specifics, and by day's end Joe had responded with a detailed pitch, five single-spaced pages long, illustrated with photographs from the field. It was a bit naïve and stilted in style, but cleverly calculated to hook Barr's bosses:

> Burma definitely has the elements for a great film. For starters, the country seems trapped in a time warp; once one leaves the large cities, one sees people getting around by horse- or ox-drawn carts; women walk around with large pots balanced on their heads; everybody wears the traditional Burmese garb (western-style clothes are nonexistent); crumbling temples are everywhere. The country just seems primitive.

Of course, the main thing Joe had to offer was his pioneering work: the discovery of the Burmese spitting cobra and his investigations into the clan of the snake catchers and the heavy illegal trade in snakes with China.

Barr wrote back immediately, telling Joe that it was just what the producers were looking for. By the end of the following week, Barr had gotten the green light—lightning speed for famously slow National Geographic. A team was assembled to shoot two half-hour programs in Burma. Australian producer Brian Armstrong would supervise the films; he recruited associate producer Nancy Donnelly and cameraman John Catto. They budgeted a month and a half in the country. Joe promised to start making the necessary travel arrangements and to help get permission for the crew to film in the countryside. Joe had answers for most of the producers' queries, but when they asked if he was planning to bring antivenom, in case of a snakebite, he replied, "No antivenom yet exists for my new species of cobra, as I only discovered it last year; we will simply take strong precautions to prevent a bite."

There were a lot of logistical kinks to be smoothed out. At the eleventh hour, the Burmese government demanded a huge cash payment before the film crew would be allowed in. Joe fired off a fax to a new friend in Yangon, Daw Marlar Sein Maung, the director of a travel company who was very well-connected with the generals who ran the country—particularly General Chit Swe, the former head of the Forest Ministry. Joe wrote:

> I need to ask your help with an urgent matter. The Ministry of Information is demanding that National Geographic TV pay them U.S.

$25,000 for filming permits. This is ridiculous. NG TV is not willing to pay that much and the whole TV project that I have worked so hard on is in danger of collapsing! Can Chit Swe help?

Daw Marlar met with General Swe and even more exalted officials whose names she didn't wish to disclose, and got the fee down to a sub-larcenous amount.

The original plan was for Joe to take Barr and the crew to a wildlife refuge in the remote northwest region of Burma, but by the end of May, the Ministry of Information let it be known that they were opposed to giving National Geographic permission to go there—for political reasons, Joe surmised. So he proposed as an alternative a visit to Pakokku to film the snake catchers in action, followed by a trip to Alaungdaw Kathapa National Park, which would involve a daylong trek by elephant—"definitely a filmworthy spectacle," Joe promised. The advantage of shooting there was that Joe's Burmese team would already be in residence, carrying out a field survey. Joe would bring Jens Vindum, Heidi Robeck, and a young research assistant named Kevin Wiseman. Rachel Slowinski happened to be in Xi'an, China, documenting traditional paper-cut artists on a grant from the Durfee Foundation; Joe persuaded her to join them.

Brian Armstrong was delighted by Joe's performance on camera: "As a guest scientist, Joe was as good as they get." Recalling his apprentice days, Armstrong elaborated, "In the newsroom, we used to say, 'Don't work with schoolteachers or scientists.' But Joe was far more savvy than most scientists, more television-savvy." Armstrong needed to get two programs out of the trip. "It was clear that the tattoo guys could stand on their own," he said. "So I asked what else could we build a show around: Is there some sort of mission? Some Holy Grail scientists are after?"

Joe told him, "We think there's a new species." He had heard reports of a second, much rarer type of spitting cobra in Burma that had a hood marking like that of the monocled cobra. Armstrong approved: It was a perfect story line. The program would be constructed as a buddy picture: Joe and Brady Barr on a quest for a new species of "mystery cobra," as the script put it. It would be called *Cobra Hunt*.

The experience was exhilarating for Joe, who had always loved the movies, from Bugs Bunny to highbrow world cinema. He especially admired Kurosawa; at a family Thanksgiving celebration in 1997, he insisted that everyone watch *Throne of Blood*. The only artwork on the walls of his apartment in San Francisco, aside from paintings by his parents, was a poster for John Huston's film version of *The Night of the Iguana*. Joe often talked about writing a screenplay—about a plucky, wisecracking American herpetologist in the Burmese jungle, naturally. His model was another Huston film, *The African Queen*, with its mix of authentic naturalism and romantic adventure. Charlie Allnut, Humphrey Bogart's character, had served as a role model of sorts for Joe in his adolescence. "I can make a better film than all this crap that's out there now," he would sometimes say with Allnut-esque bravado after he had had a few beers.

The other scientists on the expedition were less pleased about participating in the film. Jens Vindum was skeptical about Brady Barr's folksy herps-wrangling act and later said that having the National Geographic team on the expedition was "a pain in the ass." Like many people dragooned into serving as an extra, he resented the filmmakers' constant demands. "They made us walk up the trail over and over again, pick up the same frog over and over." Vindum was also offended by the artifices of cinema: "They sprinkled water on the stream to make it look like rain," he said disgustedly. Heidi Robeck was more outspoken. On a few occasions, the tensions between her and Armstrong and Barr flared into bitter words. In the finished film, Robeck's face never appears on screen.

It was Barr's first visit to Asia, so in the film—as in graduate school—Joe was cast in the role of elder. The film opens with scenic shots of the two herpetologists riding in pedicabs in Yangon, establishing them as a team, and continues with a collage of snapshots of them as children, grinning with triumph as they hold up snakes and turtles—two all-American golden boys, so alike they could almost be brothers. In the film, as they travel through the countryside together, Joe confidently teaches Barr the basics of the snakes of Burma as they encounter cobras, kraits, and Russell's vipers. Brady Barr said, "Joe was one of the slickest on-camera scientists I've ever worked with. I used to joke with him, 'Get out of the Academy, and come with us.' "

One night, near Mandalay, *Cobra Hunt* veered disastrously off script. In the film, Barr and Joe are road-cruising in the countryside when Joe hits the brakes—shouting his usual war cry, "Snake! Snake! Snake! Snake!"—and leaps out of the car. A strapping *Naja mandalayensis* lies coiled in the middle of the road, its hood flared in an aggressive posture. The snake lunges wildly and spits a rope of venom at Barr's face. Barr, wearing the protective goggles that are standard gear for dealing with spitting cobras, is unharmed. Joe, not wearing goggles, catches the snake with his tongs and eases it into a bag that Barr holds for him. As it slides into the folds of the dark sack, the unruly snake spits another load of venom.

They got back very late, to a dark, quiet camp. Armstrong needed a close-up of the snake; as the men sat around a picnic table over a few beers, the producer and his performers debated whether they should shoot it that night or wait till the next day. Armstrong didn't insist, but Joe and Barr said they wanted to get it over with. They got the cobra out of the bag, and John Catto quickly filmed the scene Armstrong needed.

When Joe was putting the snake back into the bag, it sank its fangs into the middle finger of his left hand—through the cloth. Joe jerked his hand back reflexively and pinched the fingertip, bringing forth a spurt of blood. He quickly called to Armstrong and said, "You better get John to get the camera, I've just been bitten."

In John Catto's footage, Joe nurses his finger and says quietly, as if to himself, "This is a bad bite. He got me good." He brushes aside the offer of a tourniquet and instead asks for a compression bandage. Looking downcast, his voice cracks when he says, "He tagged me through that bag." Joe struggles to keep his cool: He told Armstrong later that his worst nightmare was to get bitten out in the middle of nowhere. Trying to reassure himself, he tells Barr, "I'm really counting on the fact that he spit a lot of venom at us earlier." Joe hoped that the snake had expended its store of the toxin before it bit him. Barr bucks him up.

His face now composed, Joe says he will wait and see if he has been envenomed. "If I feel pain, then I'll make the decision to go to the hospital." The scientist in him is taking over. He talks the others through the effects of the venom, so they will know what to expect. He says, "If it starts to happen, you have to give me mouth-to-mouth, to try and help me work through it." He tells them that no matter what happens, they mustn't give him antivenom, as he is severely allergic. Everyone makes a conscious effort to remain as calm as possible, standard first-aid advice to retard the spread of the toxin in the bloodstream.

What happened next, however, was omitted from the film. Heidi Robeck became distraught when she found out that Joe had been bitten. Brian Armstrong reported that earlier that day he had seen her drink alcohol, chew betel nut, and take sleeping tablets and Lariam, an antimalarial drug known to trigger psychotic episodes. When she saw Joe sitting with his head bowed, cradling his bandaged hand, she shrieked, "Joe, if there's any venom in that, you're dead!"

The men were stunned, though Catto had the presence of mind to switch on his video camera, which captured the rest of the scene on tape.

Joe sharply swings his head around to face Robeck and says, almost pleading, "Why are you fighting with me, Heidi?"

She shouts back, "Because you're not protecting me from these fuckers!", and points at the National Geographic team.

Brady Barr loses his temper and yells, "What is your fucking problem?"

"You're my fucking problem, Brady," she snaps back.

It's the worst possible environment for someone who has been bitten by a snake. "This doesn't help," Joe says wearily. "This isn't helping. Everybody just calm down. We're all trying to stay calm."

Barr, glaring at Robeck, says acidly, "We don't need your pissing and moaning, Heidi."

Anguish washes across Joe's face. He is already dealing with stress on a level that few people ever experience in their lives, and now a student in his charge, his friend, is apparently experiencing a mental breakdown of some kind. Rawly, he says, "There's enough tension here. My fault, my fault." The others lapse into edgy silence. Joe takes the burden of changing the subject and says awkwardly, "Brady, I've caught several cobras in my life, and I've never been bitten."

In the finished film, the scene shifts directly from the snakebite to Joe sitting alone in ghostly blue darkness, anxiety etched on his face: no cinematic artifice there. Even without the awful scene with Robeck, it was a terrifying hour or two for Joe. Barr said later, "I had nightmares about that for a long time."

After a few hours, Joe still felt no ill effects, no tingling or numbness in the muscles of his arm. It had been a dry bite. His hopeful scenario—that the snake had expended its venom in its spitting behavior earlier in the evening—had proved to be correct. On camera, Joe says earnestly, "We'll be more careful the rest of the shoot."

At four in the morning, Heidi Robeck woke up Brady Barr and a few others in their tents and apologized profusely for her behavior. She got one of the Burmese staff to drive her to the airport in Mandalay and flew back to the States by herself. Joe might have been excused for not forgiving Robeck for the episode, but he did: He invited her to return to Burma with him a few months later.

In the morning, Armstrong filmed a scene of Brady Barr checking up on Joe. Barr strides into the room to find Joe packing and asks breezily, "How are you feeling?" Joe tells him he is fine. Barr asks, "No swelling? No anything?"

Joe grins in reply. Barr tells him, "You're a lucky man."

The National Geographic team headed out immediately to begin shooting *The Clan of the Snake Catchers*, in Pakokku. It turned out to be a less interesting film, but after the near-catastrophe in Mandalay, this was a disappointment to no one. Like *Cobra Hunt*, it's structured as a quest: In this case, Brady Barr is searching for a legendary, mysterious cult of "bare-handed snake busters." Of course, the quest was a contrivance: The identity and location of the snake catchers were well known to the filmmakers from the beginning. Joe, still recovering from his cobra bite, only appears in the film in a brief interview clip.

The group's stay at Alaungdaw Kathapa National Park ended with another blast of unintended drama. The park is notorious for mosquitoes and thus malaria; several people on the expedition came down with it. A Burmese porter died. The day after Joe got bitten by the cobra, when the group was on the road to Pakokku, Rachel Slowinski took sick. At nightfall they stopped at a restaurant, but Rachel felt so ill that she couldn't get out of the car. Soon she was burning up, and stumbled into the restaurant, where she collapsed.

Joe took control, saying that they had to determine how high her fever was at once. Assistant producer Nancy Donnelly found a thermometer and took Rachel's temperature: The mercury was off the end of the stick. Rachel thought she was going to die; soon afterward, she lost consciousness. Joe calmly declared, "We have to bring her temperature down." He got a bag of ice cubes from the kitchen, packed them against the back of her neck, and poured ice water all over her body.

When Rachel came to, Joe bundled her into the car and drove off to find a hospital. They soon found a squalid clinic with a lone, rickety bed, a dirt floor, and an outhouse for a toilet. The nurse on duty instantly diagnosed malaria, but she didn't have any medicine except painkillers. Joe found a single tablet of Fanzidar, a potent remedy for the disease, and gave it to her. He told Rachel, "You can't vomit this, it's the only medicine we have." Joe's prompt first aid probably saved her life.

Working with the National Geographic film team was an exciting experience for Joe. At the same time, perhaps inevitably, he was disillusioned by it. A few months later, he wrote in his journal: "I am now fairly pissed at NG TV because I feel like they got a better deal out of our interaction than I did. They did things without regard to the consequences for me." He was particularly disappointed "that they cut me out of the second film," *The Clan of the Snake Catchers*. After all, it was he who had proposed the subject. Yet Joe was always derisively unsympathetic to complainers, so he had to be a good sport about it. Anyway, he wanted to do it again.

When he got back to San Francisco, Joe found that he wasn't the only one who was disenchanted. Word had gotten around the Academy about his brush with death by snakebite—on videotape. This was not the sort of publicity the California Academy of Sciences was looking for. The snakebite itself was bad enough, but if the finished film included even a glimpse of the traumatic scene with Heidi Robeck, it would be a public-relations disaster. Joe wrote the producers an uncharacteristically turgid letter, requesting that they cut the footage of his snakebite; parts of it sound as though they were dictated to him by a committee. It's worth printing in full:

> I have given considerable thought to the snakebite scene and have also discussed this with my colleagues. Unfortunately, I can see no personal good that can come from including that scene in the film. Yes, I know it makes for good TV, especially when there is a need to capture some of the market share from the Discovery Channel. But I stand to be hurt by this in ways small and large, ranging from criticism by colleagues at meetings to being denied grants. In my field, people like Brady and I who work with dangerous reptiles are viewed by some with suspicion, with the feeling that we do it because of the danger rather than the science. As long as accidents don't happen, these people are silent. But when an accident happens, we have to endure the scorn and criticism. This could easily lead to me being turned down for grants or promotions.
>
> But beyond me, there is the potential for damage at wider levels. I have talked to Jens and the other curators and they feel there is the potential for the department to suffer negative publicity. Given that we depend on donations, this is a real concern. This also applies to the entire Academy.
>
> I feel that the most important consideration here is the damage that might accrue to me and the Academy. I hope you'll understand and modify the film accordingly. Please call if you have any questions. I appreciate it.

It's a naïve letter, in many ways. Not only had Joe signed a release giving Armstrong's team full access to the activities of his expedition, he himself had told Armstrong to instruct John Catto to turn on the camera immediately after he was bitten. If Joe really expected that the filmmakers would leave out a scene that made for "good TV" just because he asked them to do so, he wasn't as media-savvy as he and Brian Armstrong thought. (It's also possible that Joe was just going through the motions and had no expectation that Armstrong would do as he asked.)

Joe's request that the snakebite scene be cut was refused, of course: As he himself had said, it made for gripping, brilliant television. While National Geographic Television had no desire to bring negative publicity to Cal Academy, protecting the museum from exposure it might not want was hardly a priority. Yet everything in the finished program was directly related to Joe's activities as a herpetologist; no mention was made of the fracas with Heidi Robeck.

Despite the appearance that Joe's letter was written under orders, there's no reason to assume that it was insincere. As stilted and vague as the letter is, the line "when an accident happens, we have to endure the scorn and criticism" suggests that Joe himself had endured such scorn personally. In fact, the letter reveals more about the psyche of the herpetologist than Joe could have intended.

A faint patina of embarrassment cloaks the derring-do of field herpetologists. Bryan Fry, the Melbourne-based scientist, said, "At heart, all herpetologists are still eight-year-olds holding a pillowcase with something squirming inside it." Fry pushed the metaphor, with a reference to the comic strip *Calvin and Hobbes*: "Calvin is out in the woods, and Hobbes asks him what he's doing. Calvin replies, 'I'm looking for frogs.' Hobbes asks why, and Calvin replies, 'I'm obeying the exhortations of my innermost soul. My mandate also includes weird bugs.' "

When Joe traveled throughout Burma, he habitually free-handled the country's most dangerous snakes, the cobras and kraits, often with a throng of local children watching. Joe was an experienced and skillful snake wrangler, but such displays were definitely not a part of standard field procedure for academic herpetologists. George Zug was on a field expedition with Joe only once, in May 2000; after that experience, Zug told his young colleague that "he should not expect me to hold a snake bag for him. Joe had professional snake-handling skills—he just didn't practice them."

Zug, like most herpetologists, strictly avoided any direct physical contact with venomous snakes. There is no formal, universal protocol governing snake-handling techniques, but the great majority of herpetologists use specialized tongs, hooks, and sticks with remote-controlled clamps for grasping the snake, which permit near-total control of the animals. Some biologists add another layer of safety by using Kevlar gloves, but pay the price of reduced dexterity. Joe's occasional daredeviltry in handling hot herps was taboo, but he was not alone in the practice: Some academic herpetologists simply can't resist the danger of direct contact with lethal reptiles, whether they admit to it or not.

All his life, Joe felt the compulsion to emulate the fearless snake wranglers he had known, going back to the Hopi priests he had seen in New Mexico when he was a boy. Yet at the same time he wanted his accomplishments as a scientist, his ingenious contributions to the higher realms of evolutionary theory, to be admired, as every creative person wishes his work to be admired. A sense of injustice that verged on the irrational pervaded that letter to National Geographic. Surely his colleagues at the California Academy of Sciences didn't react to the news of Joe's snakebite with scorn—that would have been inhuman. The rub did not really lie in the fact of his having been bitten; it was that the incident was chronicled in a film aimed at a wide audience. The guild was prepared to accept with perfect stoicism an accident befalling one of its own while in the pursuit of science, but they didn't like outsiders catching a glimpse of that eight-year-old holding a pillowcase with something squirming inside.

Inland Taipan

Oxyuranus microlepidotus

Also commonly called the "fierce snake" in the remote Australian outback where it lives, the inland taipan holds the distinction of possessing the most toxic venom of any species of snake: It's about 68 times as poisonous as that of the king cobra, and 740 times that of the western diamondback rattlesnake. One bite could kill a hundred grown men, but in fact no human deaths from the species have been recorded, owing to its extremely isolated habitat, the dry flood plains of central Australia, and the efficacy of antivenom in treating bites. The species ranges from six to nine feet in length; its coloration undergoes dramatic seasonal changes, from dark brown or olive in winter to a light straw color in summer. During the day, the inland taipan prowls in search of small mammals, especially rodents such as longhaired rats, which can reach plague proportions in the region. The snake subdues its prey with a series of rapid, accurate strikes, which inject the highly toxic venom deep into the rodent.

By 2000, Joe had established himself as a major player in international herpetology. He had just discovered and described two new species, and more were on the way; with George Zug, he had created the Myanmar Herpetological Survey, a major, self-sustaining scientific collaboration between Cal Academy, the Smithsonian, and the Burmese Forest Ministry; and his essays in theoretical phylogeny were becoming ever more influential. Now he turned his attention to U Uga's standing promise of an expedition into the Hkakabo Razi region—the far north of Burma, its toehold in the Himalayas.

Joe envisioned an ambitious expedition into this largely unexplored region with researchers in many fields, from several major international institutions. It would be a scientific safari on a scale exceeding anything ever previously attempted in the ancient forests of Upper Burma. Joe was confident of making major discoveries: In an area so poorly studied, it was as close to a certainty as science gets. Fresh from his success with the television division, Joe was working on a proposal for an exploration grant from the National Geographic Society to pay for what would be an expensive expedition. He also hoped to lure Brian Armstrong back to Burma to make another film.

Yet Joe soon ran into some sticky political problems. His initial attempts to get permission for the expedition languished in the wasteland of the Burmese bureaucracy, and soon he ran into trouble at home, from Alan Rabinowitz. Rabinowitz had achieved international renown with his programs at the Wildlife Conservation Society to conserve the big cats of Thailand and Central America, and he exerted a powerful influence over field science in Burma. He was quite open about his opposition to the rising herpetologist at Cal Academy. Rabinowitz's attitude was "Work with Joe Slowinski, and you'll never work with me again."

The two zoologists were essentially rivals for the favors of U Khin Maung Zaw, the director of the wildlife division. Rabinowitz had given him Hkakabo Razi National Park; Joe matched that with immortality in the taxonomic literature in the lithe shape of the wolf snake *Lycodon zawi*. On his last trip to Yangon, Joe had invited Zaw to visit California; he had also given him the unrestricted use of a new Mitsubishi four-wheel-drive vehicle he had bought for the use of the survey, when the Americans were out of the country. (In fact, Zaw had already taken possession of the vehicle while Joe was in San Francisco, so it was more a matter of acquiescence.)

One of the first things Joe did after he decided to put the expedition together was to call Chris Wemmer, who in addition to being very knowledgeable about how things worked in Burma was a charming, popular guy. He advised Joe to engage Rabinowitz: "Tell Alan you want his help." So in 2000, Joe approached Rabinowitz with his best attempt at modesty, and the two met at a quiet garden restaurant near the Forest Ministry, on the outskirts of Yangon. Rabinowitz said, "Joe told me he wanted to come into Hkakabo Razi jointly with us. I told him he needed to be in better shape for a trip like that. It's all up and down. I thought he was arrogant and wanted to be famous overnight."

The two men never met again.

After Rabinowitz turned him down, Joe went over his head and appealed directly to U Khin Maung Zaw. For once, however, Joe's timing was bad. Rabinowitz was having troubles of his own with the Burmese, because of a helicopter trip in the Hkakabo Razi region that had gone dangerously awry. In Rabinowitz's version of events, recounted in his book *Beyond the Last Village*, the pilot charged with transporting Rabinowitz to a remote location in Upper Burma took a wrong turn and nearly crashed in a stone canyon, then landed far from the appointed destination, where he ordered Rabinowitz to get out. After his return, the pilot lied about his blunder to his commander and put all the blame on the pushy American scientist. He claimed that Rabinowitz had disembarked against orders. Burma's minister of defense was furious when he heard this, and ordered Rabinowitz and his colleagues out of the area because they had "walked off into the forest on [their] own."

The forestry minister toed the line and declared a moratorium on all travel by foreigners to protected areas.

Joe would have to find another way to get to Hkakabo Razi.

As his career became more demanding and he spent more and more time in Burma, Joe ended his relationship with Ashleigh. Later she said bluntly, "He dumped me over the phone." A week afterward, when Joe's father was in San Francisco on a visit, the two men discussed it over dinner at a Swiss restaurant in the Sunset district. Joe admitted that he had bungled the relationship. Later that night, he wrote in his journal: "The problems with Ashleigh and me were simple. I wouldn't put Ashleigh and the relationship at the front of my life, and her resentment of that sparked the fights that I grew to hate."

After *Cobra Hunt* was broadcast, Joe's family and friends were proud of him, but they were also aghast at the snakebite scene. When Joe saw Stan Rasmussen on a visit to Kansas City, the two men shot a few racks of pool at one of their old haunts, reminiscing about the days when they chased copperheads and rattlesnakes in the woods around the Kaw River. On the drive home, Rasmussen finally scolded him: "You dodged the bullet this time, but what about the next?" Joe admitted to his old friend that the incident had scared him, and he vowed to be more careful in the future.

Joe's fear that he might lose professional prestige as a result of the broadcast of *Cobra Hunt* proved groundless. In 2000, he served as an advisor and public spokesman for a living exhibition at Cal Academy called *Venoms: Striking Beauties*. It was the largest show created at the museum in more than a decade, with all sorts of lethal creatures caged behind Plexiglas. In addition to cobras, kraits, and rattlesnakes, there were death stalker scorpions, giant venomous centipedes from Vietnam, black widow spiders, velvet ants, a poison-dart frog, fire worms, sea anemones, and blue-ringed octopus—all of them equipped by nature to deal death, if not to humans then to smaller prey. A video monitor at the exhibition played a tape of *Cobra Hunt* continuously. The show was a hit: The crowds were so big that the Academy extended it into 2001.

Indeed, after the National Geographic film aired, Joe became something of a public figure in the Bay Area. On the day that the *Striking Beauties* exhibit opened, the *San Francisco Chronicle* ran a profile of him, with Dong Lin's iconic portrait of Joe looking very dashing in a moustache and dark aviator sunglasses—with a Burmese spitting cobra, hooded and erect, in front of him. His self-confidence bolstered, Joe was now actively seeking mass markets: He was always exchanging speculative e-mails with natural-history filmmakers. Often, too, Dong Lin acted on his behalf. The week after the National Geographic broadcast, Lin was on the phone with the *San Francisco Examiner,* the Bay Area's afternoon newspaper, prodding them to give Joe some coverage. In June, the *Examiner's* Sunday magazine put Joe's article on the cover, with a splendid Dong Lin photograph of the author gliding through the jungle on an elephant.

Joe wrote a pair of articles for *California Wild*, the museum's magazine, to support the venoms exhibit. One was a field diary of his expeditions to Burma, with a fascinating gallery of Lin's photographs. The piece focused on the discovery of *Naja mandalayensis*. He told about the time one of the newly discovered species spit at him:

> I wear protective sunglasses, but suddenly the cobra darts between my legs, looks up, and lets loose with a barrage of venom. Because of the angle, some of the venom hits my eyes. I feel an instant, immense burning pain, especially in my left eye, which has taken most of the venom. I start to walk to the water jug, but already my vision is so blurred that I cannot see where I'm going. I yell for water. Within seconds, Dong is pouring water liberally into both of my eyes.
>
> Our driver suggests that I try a local folk remedy. I am laid on a bench, and the juice from tamarind leaves is squeezed into my eyes. More searing pain! This hurts even worse than the venom. I bolt upright, yell in agony, and pour more water into my eyes, which are now ruby red. I lie down and try to relax. After several hours pass,

my vision is fine, and I feel completely recovered. Did the tamarind juice work? I don't know. If tamarind juice is not available, I'm told, lime juice works as well.

The other article was a sophisticated introduction to the biochemistry and pharmacology of snake venom. Joe compiled a list of the ten most toxic snakes, as measured by the LD-50 standard, which expresses the amount of venom in milligrams required to kill mice weighing one-half kilogram (or, more correctly, a lethal dose for 50 percent of a one-kilogram sample).

SCIENTIFIC NAME	COMMON NAME	RANGE	LD-50(MG/KG)
1. *Oxyuranus microlepidotus*	Inland taipan	Australia	0.025
2. *Pseudonaja textilis*	Eastern brown snake	Australia	0.0365
3. *Aipysurus duboisii*	Dubois' sea snake	Australia	0.044
4. *Pelamis platurus*	Yellow-bellied sea snake	Indian & Pacific oceans	0.067
5. *Acalyptophis peronii*	Horned sea snake	Tropical Asia & Australia	0.079
6. *Oxyuranus scutellatus*	Coastal taipan	Australia	0.106
7. *Bungarus multicinctus*	Many-banded krait	India & Southeast Asia	0.108
8. *Hydrophis melanosoma*	Black-banded sea snake	Australia & Southeast Asia	0.111
9. *Enhydrina schistosa*	Beaked sea snake	Indian & Pacific oceans	0.1125
10. *Boulengerina christyi*	Congo water cobra	Africa	0.12

There are many such lists, and none of them is quite the same: Some species have been reclassified by herpetologists, and measurements of toxicity vary among individuals of a species. Yet every list puts the inland taipan, *Oxyuranus microlepidotus*, at the top. Most of the snakes on Joe's list are found in Australia, part of the evolutionary heritage of the island continent's geographical isolation. The deadliest land snake outside Australia is the many-banded krait.

Lethality of venom isn't the sole measure of danger to life, and mice aren't miniature human beings: There's no scientific way to correlate the LD-50 standard to the effects of a particular venom on people. In his article, Joe cited three further criteria for assessing a snake's danger: fang length, size of the venom gland, and aggressiveness. In the first category, the Gaboon viper of equatorial Africa rules, with fangs up to two inches long. The species with the largest venom gland is probably America's eastern diamondback rattlesnake, which can produce and store up to ten milliliters of toxin. While there's no objective measure of aggressiveness, the Central American bushmaster, the black mamba of Africa, and the king cobra all have reputations for being irritable and quick to attack. Joe singled out the Russell's viper as a particularly belligerent species, always ready to lash out at human intruders.

Bryan Fry has chased down and caught many thousands of these deadly serpents, in part to study their venom. "Working with some of these snakes is the biggest adrenaline rush you could ever have," said Fry. "I used to do extreme ski jumping and big-wave surfing, but none of that can touch working with these animals." He has made a particular study of the inland taipan: He and his colleagues have identified and patented a molecule of the species' venom, which may prove to be useful as a treatment for congestive heart failure.

In the summer of 2000, Joe shifted north, into China, where he led an expedition to the Gaoligongshan Mountains in Yunnan, just across the border from Hkakabo Razi. Ten scientists from Cal Academy and an equal number of Chinese researchers followed the Nujiang River, as the Salween is called in Yunnan, into a land of clouds—storm clouds, as it turned out. The rain was heavy and continuous, causing floods and mudslides, which frequently shut down the roads until a bulldozer could be found to clear them.

The people who live in these mountains call the Nujiang "the angry river." At a steep gorge a few miles from the border with Burma, the scientists saw, far below them, the corpse of a woman caught in a ferocious eddy surrounded by sharp rocks. She had fallen there days ago, but no one had been able to retrieve her bobbing, bloated body, the arms extended upward as if beseeching help. It was an eerie, disturbing image that no one on the expedition would soon forget.

Following the pattern of his Burma project, Joe formed an alliance with the province's major biological research institution, the Kunming Institute of Zoology. He also recruited Rao Dingqi, a young herpetologist on the staff there, to come on the Hkakabo Razi expedition. To raise funds for the Yunnan trip, Joe and Cal Academy botanist Peter Fritsch wrote a major grant proposal for the National Science Foundation, to fund a biodiversity study similar to the Myanmar Herpetological Survey.

On his next trip to Burma, Joe brought along zoologist Douglas Long, who would also join the Hkakabo Razi expedition in 2001. On the way, they made a quick trip to Laos, a country just as remote and understudied as Burma, to assess the possibility of organizing an expedition there. They saw spitting cobras, and Long said that he counted 120 species of birds in four days. They visited the Plain of Jars, a grassy plateau in northern Laos strewn with hundreds of mysterious stone urns from three to ten feet high, which French archaeologists believe were used as ossuaries by a vanished Bronze Age civilization. (During the war in Vietnam, American B-52s flew more than 63,000 bombing sorties over the Plain of Jars, turning the landscape for many miles into a wasteland of rubble—but sparing the jars.) Long would later recall of their visit to the Plain of Jars only that he and Joe got "extremely tanked on *lao-lao*," the local white lightning. Joe never returned to Laos.

For the first time in his professional life, Joe Slowinski was being sought after. Biologists from St. Petersburg to India to Venezuela wanted to collaborate with him. He was even courted for an open curator's position by the American Museum of Natural History, perhaps the most prestigious institution for biological field research in the United States. Despite the fact that the salary would have been substantially higher in New York, Joe wasn't interested in leaving San Francisco. He replied, "No. I am extremely happy at Cal Academy. Great people, great city, great food, great friends." He said he didn't want to leave Jens Vindum, the collections manager at the Academy: "We click together so well. I consider him one of the best people to work with in this field." It was just as everyone had said when Cal Academy hired him: It was the perfect job for Joe.

As 2000 drew to a close, Joe solved the major obstacle to launching a scientific expedition to Hkakabo Razi by giving up altogether on U Khin Maung Zaw and the Forest Ministry. Instead, he went to Daw Marlar, his well-connected friend in the travel business, who got permission for the group from the Ministry of Hotels and Tourism. Tourism, with its irresistible lure of foreign exchange, was just about the only legitimate business in corrupt Burma that had achieved a reasonable level of development. Joe's strategy was unorthodox but not without precedent, and he made no secret about it; U Khin Maung Zaw even gave tacit approval to the plan by granting permission to Joe's field team from the wildlife division to accompany the expedition on an unofficial footing. One advantage to working with a crooked, autocratic government was that there were few rules that couldn't be bent, if you knew the right people.

Then came a major setback: Joe's application for funding from the National Geographic Society was rejected. Of course, the Society turned down many fine proposals and never explained why; even an institution as rich as National Geographic didn't have enough money to pay for every worthwhile research project that came begging at its door. Yet Joe was convinced that Alan Rabinowitz was responsible for the rejection.

It was even worse than that: He believed that Rabinowitz was going around telling anyone who would listen that he, Joe, had raised his expedition without obtaining the necessary legal permits and—most offensive of all—that the Cal Academy group's collecting activities would pose a threat to the biodiversity of the region. When these reports reached Patrick Kociolek, the executive director of the California Academy of Sciences, he called Joe into his office to reprimand him. Shattered by the experience, Joe phoned his father afterward to vent his frustration, saying that Kociolek had sternly told him, "I'm so angry I want to sack you: How can you take my scientists to Burma without the proper permits?" Joe set him straight, explaining the unusual arrangements he had made, but it was a close call.

The injustice of the scenario infuriated him; he played it over and over again in his mind. No one was more dedicated to the preservation of biodiversity than he was. Why was this guy Rabinowitz so hostile to him? Weren't they on the same side? It seemed to Joe that Alan Rabinowitz was defending his turf in Burma with all the tenacity of the big cats he had done so much to help preserve. Park or no park, Joe thought he had a right to visit Hkakabo Razi: U Uga had invited him to mount an expedition there on many occasions. Other scientists in the field complained about Rabinowitz's heavy-handed tactics behind his back, but they were afraid to challenge him: He had too much clout. Again, Joe's frustration was intensified by the conviction that herpetologists get treated as second-class citizens by Big Biology. Rabinowitz was a champion of the noble feline; Joe, who studied the creeping, dust-eating serpent, felt that he would never get his just deserts from the establishment.

Rabinowitz, for his part, thought he had the prerogative to monitor the foreign scientists who wanted to carry out research in Hkakabo Razi National Park. He was working out a management plan to administer the park following international guidelines and was offended that Joe wanted to go there before anyone else. Despite his personal objections to collecting wildlife, Rabinowitz said he would not have opposed Joe's expedition if it had confined its route to wilderness areas outside the boundaries of Hkakabo Razi National Park. The influential mammalogist, it was true, assumed a certain proprietary air about the new preserve, but he was only following tradition. An unwritten law of science holds that one researcher doesn't venture into another's domain without permission. Rabinowitz had created Hkakabo Razi National Park, and even if the legal responsibility for it now lay with the Forest Ministry, he felt that he had the moral authority to decide who should be allowed to work there, and when, at least until the park was well established.

Joe's suspicions were justified. When the Forest Ministry asked Alan Rabinowitz about Joe's application to take an expedition to Hkakabo Razi, he replied, "I do not recommend that Joe Slowinski be given a permit." Rabinowitz had undermined him at National Geographic as well. "When I found out about Joe Slowinski's expedition," said Rabinowitz, "I was working with National Geographic on a special about the jaguar in Central America. Joe told them that he had the proper permits, but I told them no, he didn't have Forest Ministry permits. I simply told them that if they supported an expedition that was doing the wrong thing in Myanmar, then I would not work with them."

A grants officer at National Geographic called Chris Wemmer, a neutral observer, and said, "We have a proposal here from Joe Slowinski, but there has been one highly critical review. What's going on?" Wemmer told him, "It's all about ego." Wemmer did what he could to explain Joe's position without injecting his own opinion. But Wemmer felt that Rabinowitz was acting "like a spoiled kid—it was 'me or him.' Alan didn't want any threats to his professional supremacy."

Feuds among field biologists have a long, eccentric pedigree. America's greatest herpetologist, Edward Drinker Cope (1840–1897), carried on one of the bitterest, most famous vendettas in the history of modern science with a paleontologist from Yale, Othniel Charles Marsh. Cope identified and named an astounding 1,282 genera and species, 510 of them amphibians and reptiles. In the years after the Civil War, the

Morrison Formation, the vast rock field centered in Colorado and Wyoming, was beginning to disclose its hoard of vertebrate fossils, including some of the best dinosaur remains ever excavated. Hundreds of them—many still on display more than a century later in public collections such as the Smithsonian—were described and named by Cope.

And most of the rest were studied by O. C. Marsh. The two began as friends, but quarreled after Cope took Marsh on a tour of his best collecting sites in New Jersey, and returned the next summer to find that Marsh had secretly bought the mineral rights to the land. Young Cope went west, where the fossils were more complete and in better condition; Marsh followed him. Their territorial squabbles soon escalated into a titanic duel of egos. The two men worked at a furious pace side by side, sometimes rushing papers into print within days of each other. It was inevitable that errors would creep in.

Marsh caught Cope in a really bad one. In Cope's description of Elasmosaurus, a giant plesiosaur, published in 1868, he wrote that "propulsion was more largely accomplished by its tail than by its paddles"—but the "tail" he was referring to was actually Elasmosaurus's neck: Cope had attached the great reptile's skull to the end of its forty-foot tail. After being publicly humiliated by Marsh over this blunder, Cope was determined to have his revenge, so he enlisted Marsh's chief assistant to work for him as a spy. As the feud escalated, the competing crews dynamited their own and each other's sites to keep their opponents from excavating, and even stole fossils from each other. The *New York Herald*, the country's leading newspaper at the time, made the quarrel between the country's foremost biologists front-page news, turning it into a tawdry beard-pulling spectacle, which dragged their respective institutions and academies into the fray.

Their rivalry ended only with Edward Drinker Cope's death in 1897, at the age of fifty-six. He belonged to the Anthropometric Society, an organization of geniuses straight out of Jules Verne, who basically wanted to study why they were so smart. They pledged to leave their brains to science; Cope's was preserved in formalin at the University of Pennsylvania. Before his death, Cope had proposed that he himself be declared the type specimen for *Homo sapiens*—the only known species on Earth that has no holotype. In 1994, the modern paleontologist Robert Bakker revived Cope's extravagant proposition, carefully measuring what was left of the great man's remains and publishing the results in a scholarly journal, tongue firmly in cheek.

Another celebrated feud took place closer to home for Joe Slowinski, in the ranks of the herpetology department of the California Academy of Sciences. One of Joe's most distinguished precursors, John Van Denburgh, the curator who rescued the holotype specimens from the fire of 1906, had an acrimonious feud with an assistant, a medical doctor named Joseph Thompson, who had collected some twelve thousand Asiatic reptile specimens for the Academy. The point of contention was priority in publishing new genera and species of herps collected in Taiwan and the Ryukyu Islands, an archipelago between Taiwan and Japan. In 1912, within a few days of each other, the two men issued nearly identical pamphlets, giving the same names to the new taxa. On the title page of his pamphlet, Van Denburgh identified himself as "a writer who has long been known most favorably to all herpetologists, the world over, as an accurate, painstaking, and skillful scientist."

Thomas Barbour, an influential curator at Harvard's Museum of Comparative Zoology, took it upon himself to straighten the mess out and wrote both men polite letters of enquiry. Joseph Thompson's reply to Barbour began: "Your letter of December 14th is written in a grossly insinuating tone. It is obnoxious to the limit. The alternative presents of replying in terms that would border on being contraband in the mails, or of calling attention to the absurdity of your attempting to sit in judgment on matters which your distance from the scene of action and ignorance of the facts combine to prevent the forming of an intelligent opinion." Thompson went on to accuse Van Denburgh of "the old, old trick of rushing into print with material belonging to another."

Van Denburgh's reply was more temperate in tone, but full of dark accusations. He claimed that Thompson had copied all of his notes and "then proposed that I should give him joint authorship in all papers I should ever write . . . whether or not he had anything to do with their preparation." When Van Denburgh refused, Thompson broke off all contact and threatened to take his reptiles back. The Academy subsequently held an investigation and found in Van Denburgh's favor.

Joe Slowinski's ego was modest compared with those of Edward Drinker Cope and John Van Denburgh, but once he had made up his mind to do something, nothing could deter him. When National Geographic refused Joe's grant, nearly two years in the planning, he was undaunted: He knew he could find the money from other sources. Scarcely missing a beat, Joe instructed Daw Marlar and her company, Thurein Travel, to create a detailed itinerary and final budget for the expedition. He had firm commitments from a good group of scientists: In addition to the core group from the Academy, embracing many disciplines, he had recruited Berkeley entomologist Mark Moffett, Harvard botanist David Boufford, and two biologists from Yunnan. This party would not be canceled.

Chris Wemmer said, "I knew that if Joe didn't get the grant, he would persevere. He had already made his mark; Alan Rabinowitz couldn't make him go away." Wemmer also recognized Joe's charisma: "As a snake wrangler, he was every bit as good or better than the snake chasers on television. He could have been making millions doing the same thing, but he wasn't into self-promotion." In Wemmer's view, Joe never got the credit he deserved. "The irony is that the crusader who came out of all this, the hero, was Joe Slowinski."

A month before he left on the "big expedition," as he always called it, Joe gave a lecture at Bohemian Grove, the elite, highly secretive men's summer camp in Sonoma County, at the invitation of Robert Drewes, the chairman of the herpetology department at the Academy. A cross between Davos and *Animal House*, Bohemian Grove is located in 2,700 acres of ancient redwood forest, about sixty miles north of San Francisco. The campsite is owned by the Bohemian Club, the most exclusive men's club in California. It was founded in 1872 as a drinking club by a group of journalists (Mark Twain, Jack London, and Bret Harte were among its early members), but it evolved into a traditional gentlemen's club for the city's establishment. The camp in Sonoma was where the movers and shakers went every year for an interlude of Dionysiac abandon, free from the constraints of wives, shareholders, and reporters.

By the 1920s, Bohemian Grove was attracting power brokers from all over the country. Herbert Hoover, who called it "the greatest men's party on Earth," was drafted to run for president there, and every Republican president since Hoover has attended. The year before Joe visited, George W. Bush announced his choice of Dick Cheney as his running mate after the two emerged from the encampment. (Richard Nixon hated it: In 1971, he told John Ehrlichman and H. R. Haldeman, "The Bohemian Grove—which I attend, from time to time—it is the most faggy goddamned thing you could ever imagine, with that San Francisco crowd.") For years it has been a coveted invitation for the global elite. When Lee Kuan Yew, the creator of modern Singapore, answered the call, he was mistaken for a waiter by other campers.

Bohemian Grove used to have a riotous reputation, but in recent years the outrageousness seems to be confined to gin fizzes for breakfast and urinating on the redwoods. The opening-night ceremony is an elaborate ritual that culminates in the burning of an effigy of Care on an altar before a forty-foot-tall statue of an owl. Joe was too hip to be very impressed by the creaky hocus-pocus, but ambitious enough to be awed by the confluence of power at the camp. Other lecturers talked about Shackleton's Antarctic expeditions, how birds evolved from dinosaurs, and the future of the Pacific salmon. In Joe's talk, which was entitled "Adventures with Snakes in Burma, or Why I Wear Sunglasses" (the subtitle being a reference to his encounter with the spitting cobra), he explained that his passion for the country had begun long before he set foot in it. "It all started when I was a young kid," he told the campers. "For some reason, I specialized in old stamps from the British Empire. I remember being tantalized by the exotic images on old Burmese stamps: tigers and elephants, precious gems like rubies and jade, and amazing peoples like the giraffe women, who have a series of brass rings around their necks."

In August, just weeks before he left for the big expedition, Joe experienced two life-changing events. The first was that the National Science Foundation, which already sponsored his research in Burma, awarded him and Peter Fritsch a $2.4 million grant to study biodiversity in Yunnan. It was the largest public research grant in the 148-year history of the California Academy of Sciences. Previously, Joe had been appointed to take over as chair of the Cal Academy herpetology department after his return from Burma. Joe Slowinski's position as a leader in his field was now unquestioned.

The other event was even sweeter: He fell in love. Sandy Scoggin was doing ornithological research for a wetlands conservation group in Marin County. She was gentle, sensitive, a few years younger than Joe, and she loved the outdoors. They met at a party Doug Long threw for the scientists who were going to Hkakabo Razi. Sandy came with Maureen Flannery, her housemate, who was on the expedition roster as a research assistant in ornithology. Moe, as she was known, was nervous and excited about her first major expedition, and wanted Sandy to meet the people she was going to trek through the jungle with—especially the famous Joe Slowinski.

It was a luau theme party, in Long's eccentric tropical garden of a backyard in Oakland. Joe, wearing a Hawaiian shirt, was behind the bar, standing by an aviary. Sandy told him she was fascinated by cobras—the best thing she could have said if she wanted to catch Joe Slowinski's attention. He made her a strong gin and tonic and turned on the charm. He promised to send her a photograph of a black African cobra. "I don't know if it was love at first sight," Sandy said, "but it was certainly interest at first sight. My stomach felt all funny." A slim, tan brunette, Sandy was just Joe's type. At work on Monday morning, he sent a message by roundabout means, telling Moe Flannery, "God, Sandy is so sexy."

For their first date, a few days later, Sandy came to the Academy; she met Joe in front of a dinosaur. He gave her a tour of the museum, took her to an Irish pub to drink beer, then Japanese dinner and back to his place to watch videos (including *Cobra Hunt*, of course). The second date was an oyster lunch followed by a rattlesnake hunt in rural Marin

County. As they were hiking in a protected area, Joe veered off the trail into the rough, where visitors weren't supposed to go, and found a rattlesnake under a rock, much to his satisfaction.

It was clear where things were headed with Sandy, but Joe was taking his time; he didn't want to spoil it. He was touched by her in a way he never had been before. One Friday night, he invited her to his place for dinner. He cooked his famous shrimp flambé—which, embarrassingly, failed to catch fire. They drank a bottle of wine and watched *The Night of the Iguana*. Joe wouldn't sit next to her, so Sandy made the first move. She asked, "Are you a little bit shy?" He said yes, so she said, "I really want to kiss you." Then Joe lost his shyness.

The lovers saw as much of each other as they could before Joe had to leave for Burma. He came out a few times to visit the funky old beach house in Bolinas that Sandy shared with Moe Flannery and another friend, a biologist named Sue Abbott. The Friday before his departure, Joe spoke about the upcoming expedition at the weekly beer social at the Point Reyes Bird Observatory, where Moe Flannery worked. Dong Lin and Doug Long came out for the event. It was the same day that Joe got the news about his National Science Foundation grant, so there was a lot to celebrate. Joe spoke more eloquently than usual. Afterward, there was a barbecue at the house in Bolinas, with dance music blasting in the crisp ocean breeze. It was a perfect night.

The evening before he left for the big expedition, Joe called Rachel to say good-bye. She teasingly scolded him for going to Hkakabo Razi without her: The two of them had often talked of going there together. She also nagged him about taking his malaria medication, reminding him of the time she almost died in Burma, and made him promise to take a doctor along.

In the morning, Sandy took time off from work so she could go to the airport with Joe and Moe Flannery. Sue Abbott drove them. Sandy was in the backseat with Joe, counting out $40,000 in cash, the expedition's treasury, and stuffing it into envelopes. At the airport, Joe hopped out with Flannery. He gave Sandy a farewell kiss and promised to send her an e-mail as soon as he arrived in Yangon.

The Expedition

On August 23, 2001, after two years of preparation for his journey to Hkakabo Razi, Joe wrote the first entry in a new journal: "This is the start of the big expedition." He was always excited at the beginning of a trip, and this one would be the most ambitious of his career. Before they left San Francisco, Joe told his teammates, "This is probably going to be one of the hardest things you have ever done." Most of the scientists from Cal Academy were flying over together on Northwest flight 27, San Francisco to Bangkok, to catch a connecting flight on a regional carrier to Yangon.

After cramming themselves into the crowded economy-class cabin, the scientists turned the marathon flight into a mile-high party. Joe sat next to Doug Long; his graduate student, Guinevere Wogan, the frog specialist, was nearby. It was her second trip to Burma. In April, she had accompanied Joe on an expedition to an elephant sanctuary on the country's west coast, near the border with Bangladesh, on which they had discovered two new species of bent-toed geckos, a new toad (*Bufo crocus*), and several new frog species.

The group drank all the free liquor and beer they could charm out of the crew until they became too loud and got cut off. Then Joe persuaded Wogan to buy a bottle of Jack Daniel's from the duty-free cart, which was then opened and passed around surreptitiously. Christiaan Klieger and Dong Lin, seated on the opposite side of the plane, came and joined the party, loitering in the aisle.

By the time the cabin crew caught on, the whiskey was almost gone. A fierce female attendant shooed Lin and Klieger back to their seats and confiscated the bottle. She told Joe, "I used to be a police officer. I know how to handle people like you."

Also on board were ichthyologist David Catania, a quiet, genial presence who habitually kept his own counsel, and Moe Flannery, at twenty-nine the baby of the group. The other scientists joining the expedition would travel to Burma separately over the next few days.

Someone had brought an advance copy of Alan Rabinowitz's new book, *Beyond the Last Village*, about his work in Burma. The scientists mockingly read choice bits of it aloud. The loudest howls of indignation were provoked by Rabinowitz's complaints about "two well-known American museums" that "collect wildlife from some of the few national parks and wildlife sanctuaries in Myanmar." He was obviously referring to the California Academy of Sciences and the Smithsonian Institution, co-sponsors of the Myanmar Herpetological Survey. Rabinowitz called them "biological and cultural mausoleums."

The party ended before dawn, when the plane arrived in Tokyo for a layover at Narita Airport. By the time the scientists arrived in steamy Bangkok, they were so bleary and hungover they didn't bother to go into the city before their connecting flight to Burma as they had planned.

Yet from the moment he arrived in Yangon, Joe was up and running with his usual ferocious energy. A reception committee was waiting for the Americans at the airport: a group from Thurein Travel, Daw Marlar's company, which was handling the ground arrangements for the expedition, and the Burmese scientists from the wildlife division, who would accompany the Americans. Among them were three young herpetologists: intense, headstrong Htun Win, Joe's principal protégé and the leader of the field team; Aung Khwi Sein, a good-looking young man with a shock of coal-black hair; and placid, stocky Pocho Soe Lwin.

A little caravan of automobiles swept the scientists and their hosts into the most depressed, run-down capital in Southeast Asia. Rusted-out cars and ancient, doorless trucks splattered mud on the broken sidewalks and shabby concrete storefronts that lined the road into town. Occasionally the mildewed wreck of a noble, colonial-era building rose above the surrounding slums, a relic of the days when the city was one of the loveliest in Asia. The cars rattled past long red signs on the side of the road, which proclaimed the People's Desire, in Burmese and English: OPPOSE THOSE RELYING ON EXTERNAL ELEMENTS ACTING AS STOOGES HOLDING NEGATIVE VIEWS and OPPOSE FOREIGN NATIONALS INTERFERING IN THE INTERNAL AFFAIRS OF THE STATE.

When Joe arrived at his hotel, it felt like a homecoming. On his recent visits to Yangon, with his fortunes on the rise, he had moved up from the Highland Lodge to the MiCasa Hotel, a comfortable, full-service Spanish posada with the ethos of an American business hotel, which had miraculously materialized on the edge of the city. The staff welcomed him back familiarly. The rooms at the MiCasa were equipped with efficiency kitchens, so that he could whip up a pasta dinner; downstairs there was a cozy bar with a pool table. He and Dong Lin would stay there; the other scientists were booked at the Highland Lodge.

Soon after he checked in, Joe stopped by the hotel's plush, cool business center to use the Internet, and dashed off an e-mail to Sandy:

> *Hi Sandy,*
> *We're alive and doing well. Lots of rain and very humid. On the 31st we'll head up to the north. Today we'll run around as tourists and I will take Moe to have some snake for lunch. Thinking about you. More later.*
>
> *Cheers,*
> *Joe*

The first order of business was to change money on the black market. That was essential, for the official rate of exchange for the Burmese kyat was six to the U.S. dollar, whereas the street rate was more than eight hundred kyat to the dollar. Only diplomats changed money at the bank. Deep in the fragrant maze of Bogyoke Aung San market, amid nattering merchants vending silk and lacquerware and cheroots, Joe exchanged a wad of hundred-dollar bills for a stack of worn, filthy kyat notes that stood four feet high; these he would keep in his room at the MiCasa. After lunch in town at Trader's, a fancy business hotel, the group visited Botataung Pagoda, famed as the home of a population of holy turtles. Unfortunately, the turtles' ponds had been drained. The creatures were now cooped up in a godown full of Buddha images, which had become dank and slimy with reptilian excrement.

Joe returned to the hotel for a basketball date with a Burmese friend, who didn't show up. It was just as well: A heavy rain began to fall in the afternoon, and it hadn't let up by nightfall. As he updated his journal in his room, trying to stay awake, Joe glumly wondered what the heavy rains might be doing to the dirt trails he and his team would soon be climbing in the country's far north.

Monday morning began with a series of meetings that brought the looming expedition into sharper focus. First, Joe met with U Khin Maung Zaw at the Forest Ministry, in a dim, dusty office lined with glass-doored cabinets full of scholarly books and old maps. Zaw was a courtly, soft-spoken zoologist with a serious interest in conservation. He was as welcoming as he could be, considering that he wasn't the expedition's host. And why would he not welcome Joe? The Myanmar Herpetological Survey thus far had been a great boon for the Burmese. Joe was giving his colleagues at the wildlife division world-class training and a new natural-history museum, at Hlawga Wildlife Park, near Yangon, to practice their skills; eventually, the team expected to expand the number of known reptile species in Burma from three hundred and fifty, where it stood when Joe's research officially began in 1998, to more than five hundred—an extraordinary advance in knowledge.

Yet in his indirect, Asian way, Zaw laid it on the line: Since the Forest

Ministry (that is, Zaw himself) had refused Joe's original request for official sponsorship of his expedition, the ministry could now offer only limited support. The field team—Htun Win, Aung Khwi Sein, and Pocho Soe Lwin, the herpetologists who had met the San Francisco group at the airport, and a scientific photographer named Hla Tun—would be allowed to join Joe's expedition on the bureaucratic fiction that the Burmese scientists and the Cal Academy group were on separate ventures that happened to be following the same route.

Another major item on the agenda of their meeting was the planned visit to San Francisco by some of the Burmese scientists, including Zaw himself, which was scheduled to take place immediately after the expedition concluded, in the first week of November. They were waiting for final approval; the generals who ran the country were reluctant to let intellectuals go abroad, fearing—with good reason—that they wouldn't return.

One subject that definitely didn't come up in his meeting with U Khin Maung Zaw (and which he had surely avoided when Patrick Kociolek called him on the carpet about his unorthodox arrangements) was the precise route of the expedition. When Zaw first agreed to Joe's plan, he nevertheless specifically forbade the Cal Academy group to enter Hkakabo Razi National Park, following Alan Rabinowitz's advice. Yet the itinerary Joe had created with Thurein Travel took him into the heart of the protected area. Daw Marlar had her own connections high in the government and was able to get the necessary permits, circumventing the Forest Ministry entirely. Joe risked alienating Zaw by flouting his prohibition to go to Hkakabo Razi, but he thought he could get away with it. The collaboration with Cal Academy was a good thing for the Burmese, and Joe was betting that Zaw would save face by looking the other way. In any case, he wouldn't find out until after the expedition was over. It was always easier to ask for forgiveness than to get permission.

Joe's next appointment was at Thurein, with Daw Marlar. A plump little woman who wore old-fashioned wire-rimmed eyeglasses, she had a soothing effect on Joe. Her English was excellent. Before she spoke, she would cock her head to one side with a tiny uptake of breath and say "Well!", with a hint of a smile, which gave her a moment to frame her words and lent a thoughtful air to even her most commonplace utterance.

They met at her office, directly across a country lane from the Highland Lodge. It was a rustic white clapboard house by a pond, pretty enough for a Currier & Ives calendar. First they went over the itinerary: On Friday morning the group would fly to Putao, the capital of the northernmost district of Kachin State, just fifty miles south of the frontier with China. After an overnight there, the expedition would begin in earnest. The group could take trucks as far as the first village on the route, Machanbaw, an old British outpost, but beyond that they would walk. Eight long marches would bring them to a village called Pangnamdim, the entry to Hkakabo Razi National Park. Beyond that, the going got tougher as the trail gradually climbed into the foothills of the Himalayas.

The plan was for the main group to stay and work for a while at Pangnamdim, while Joe, Dong Lin, and Christiaan Klieger pushed on to a village called Tehaundan, the northernmost settlement in Burma, in the shadow of the nation's great mountain. Despite U Khin Maung Zaw's prohibition, Joe planned to cut as wide a swath as possible through Hkakabo Razi National Park in his search for new herpetofauna. Lin was coming along because he always got his best pictures when he was with Joe; Christiaan Klieger's agenda in Tehaundan was to disprove a theory Alan Rabinowitz had spun about its inhabitants in *Beyond the Last Village*, which Klieger thought was wrongheaded.

On his journey to Tehaundan, Rabinowitz claimed to have met the last of the Taron, a tribe of pygmies that was on the verge of extinction. According to the zoologist, they were deliberately letting their kind die off, because they considered themselves to be ugly and mentally defective. Klieger would eventually prove that the T'rung (the native transliteration of their name, Taron being the equivalent in Burmese), far from being about to vanish as Rabinowitz claimed—"silently awaiting their extinction like the final, lonely passenger pigeon on its perch in the Cincinnati Zoo," as Klieger lampooned Rabinowitz's hypothesis in an essay published in 2005—were in fact culturally continuous with others of their kind in China and Tibet. Klieger would show that hundreds of T'rung people lived just twenty miles from Tehaundan, across the Chinese border.

Joe knew that there was little point in making a detailed, rigid itinerary for a group as large and diverse as this one, which aimed to cover long distances in territory that was new to them. It was the height of the monsoon, so the weather would be wet, bridges might be washed out, essential supplies would be hard (or impossible) to find on the trail: If you wanted to worry, there was a lot to worry about.

Weeks before he came to Burma, Joe had paid $44,000 to Thurein Travel for an inclusive package tour for the twelve visiting scientists, which he estimated was ten times the real ground costs of the trip. Most of the money would be split up as spoils by grasping government officials and military police all the way up the food chain. Daw Marlar conceded that the sum was exorbitant, but assured Joe that it covered everything. She reassured him, cooing her pat answer to every problem: "Don't *woooorry*."

Their guide was one of the most experienced men available, U Thein Aung, who went by the name of Jiro. He had been to the Hkakabo Razi region twelve times before. Daw Marlar said that her company had already sent ahead carpenters to build bamboo huts along the trail, to serve as the expedition's campsites. There would be two medical professionals accompanying the scientists, she said, at least one of them qualified to set broken bones. They would join the group in Putao.

Daw Marlar also gave Joe some inside gossip about her company. Thurein Travel had been started by General Khin Nyunt, the number-three man in the junta. He was chief of the military intelligence service, a post that had earned him a sinister reputation. Recently, another top general named Maung Aye had gotten involved in the company. Daw Marlar herself was Khin Nyunt's protégée—Christiaan Klieger vividly described her as "a courtesan of the junta"—so she wasn't sure exactly where she stood in the organization now. (Two years later, in

August 2003, Khin Nyunt became Burma's prime minister; fourteen months after that he was deposed and placed under house arrest by his rival, General Maung Aye.)

At the end of the day, Joe finally got to play some basketball, with Dong Lin and two Burmese friends. Dinner followed at the Highland Lodge. The menu was crab, Joe's favorite, and spicy mutton curry, served with plenty of iced beer. In the cozy dining room, Joe told the group about his meetings, saying that everything had gone well. Daw Marlar had made excellent arrangements for their expedition. When someone raised a potential problem, he imitated Daw Marlar and purred, "Don't *wooooorry*." It would become the expedition's catchphrase.

Later, a Thai-Burmese friend showed up, full of melancholy complaints about the terrible state of his marriage. A few consolatory rounds of beer led to a basketball challenge, Burma versus California. The stakes: cobra dinner.

In Yangon, Joe was in his element: juggling competing agendas, coping with crises great and small, and getting it all done with good humor. He could solve half the problems that came up with that disarming, gap-toothed smile of his: No one could say no to him. The scientists who were coming on the expedition to Hkakabo Razi knew that the opportunities for research would be exceptional, but they were also there, at least in part, because of him. With Joe Slowinski as team leader, the scientists had an expectation that everything would go smoothly—or as smoothly as could be expected in a poor country with a bad government. Just as important, there was the prospect of some good times.

Over the course of the week, the rest of the group trickled into Yangon. On Tuesday afternoon, entomologist Mark Moffett arrived. Based at the Museum of Vertebrate Zoology at Berkeley, he was an accomplished scientific explorer in his own right, who had published stories about his discoveries, accompanied by his photographs, in *National*

Geographic and *Smithsonian* magazines. He and Joe had met just two years before, but they had become friends. On Wednesday, the last of the Cal Academy contingent turned up: Botanist Bruce Bartholomew was an editor of the *Flora of China*, the first English-language compendium of the plant life of China, projected to fill fifty volumes when completed. Bartholomew arrived with two biologists from Kunming: Rao Dingqi, the herpetologist whom Joe had met in Yunnan, and ethnobotanist Dao Zhiling. Amazingly, Rao Dingqi had just gotten his passport the morning of his flight, and his Burmese visa half an hour after that. The last scientist to arrive was Bruce Bartholomew's colleague on the editorial committee of the *Flora of China*, botanist David Boufford, from the Harvard University Herbaria.

On Tuesday morning, Joe and Dong Lin went shopping for generators, which would be needed on the trail to recharge the batteries of the laptops and cameras, both still and video, that would document the expedition. There was so much to get done before the flight to Putao on Friday: more meetings with Burmese officials, more shopping for equipment. Every day brought a new little fire to be put out. Dave Catania, the ichthyologist, had failed to bring along some supplies necessary to preserve his fish specimens: Where to find them in Yangon? Already there were signs of dissension in the ranks. Reports came to Joe that Mark Moffett had been rude to Htun Win, ordering him around in a high-handed manner unbefitting a scientific colleague. Htun Win was proud, and wouldn't submit to that sort of condescension for long.

On Wednesday morning, the first crack appeared in Daw Marlar's preparations for the expedition. Dong Lin woke Joe up at seven in the morning, excited and upset, to tell him that Daw Marlar was demanding 300,000 kyats, around $500, for the Burmese scientists' flights to Putao. Yet the $44,000 Joe had already paid her was supposed to include the travel expenses for the Burmese. (The pretext that the Americans and the Burmese field team were on separate expeditions had been abandoned.) "Crap!" wrote Joe in his journal. "What is this shit!"

It turned out that she wanted 30,000 kyats for each of the four men, not 300,000. The actual price of a flight to Putao for Burmese citizens was 20,000 kyats, a bit more than thirty dollars, but an additional 10,000 per ticket would be required for bribes, because of the late date. Joe was cross with Daw Marlar: She had said before that the $44,000 total included all the expenses for everyone. Oh, no, she demurred soothingly, that sum only covered the foreign visitors. Joe checked his records: She had definitely included the field team's airfares in the budget. It wasn't much money; it was the principle of the thing. The grubby pile of kyat in Joe's room at the MiCasa was dwindling.

Joe coped with his annoyance in the usual way, with sport. He went for a swim at the American Club, the recreational facility for the American Embassy, with Dong Lin and a Burmese friend named Pieri. A few minutes after they were in the water, they heard the siren song of a basketball bouncing. Joe jumped out of the pool to investigate. An Embassy official named Paul Daley and his fifteen-year-old son Michael were fooling around on the court. Joe challenged them to a game. Pieri was wearing thin felt sandals, so Joe and Lin took them on first. The Daleys went up seven-two quickly because the son had the drop step on Lin, but Joe and Lin finally pulled ahead and beat them.

In the second game, Lin lent Pieri his sneakers; Joe won with him, too. Then they adjourned to the club bar. Over beers, Joe told Daley about his plans. Daley asked him if he had registered with the Embassy, and Joe said no. He explained that in his previous trips to the country he had found it best to keep his distance from the American Embassy; he had heard that close contacts there raised suspicions within the paranoid Burmese government. Daley, the chief of the political-economic section at the Embassy, assured Joe that registration was confidential and a good idea, and urged him to drop by for a visit.

Joe showed up early the next morning. The American Embassy in Yangon occupied a colonial-era bank building, a moldering Neoclassical pile on a busy street in the old part of town. Joe told Daley about the expedition in greater detail and gave him a map plotting its route.

Daley sent Joe down the hall to register with the consular section, experiencing a twinge of concern. Joe was obviously experienced at travel in Burma and knew how difficult conditions could be, but northern Kachin State was even more remote than the parts of the country he had visited before. Communications and medical care were virtually nonexistent, and the trails would be in terrible condition now, in the midst of the monsoon. In 2001, rainfall was 20 percent above the soggy average.

After his visit to the Embassy, Joe went for a solo lunch at the Japanese restaurant at Trader's. He loved sushi, and this would be his last chance for Japanese food for a long time. In the afternoon, he returned to Daw Marlar's office to meet Jiro, the guide who was charged with getting the team safely to Hkakabo Razi. An easygoing fellow with a high forehead and round, smiling face, Jiro first explained his strange, un-Burmese nickname: It had been bestowed on him by a Japanese client who couldn't say his real name, U Thein Aung. Then they got directly to the main subject of the meeting: overweight baggage. How could they squeeze as much of their equipment aboard the flight as possible without making a serious dent in Joe's stash of Burmese money?

Jiro was smart and well-read, and he lacked the shy reticence of most Burmese people. An employee of Myanmar Tours and Travel, a government agency, he held a post in the Burmese Army, but he didn't hew to the government line, as most Burmese minders did. He proved to be practical and plainspoken about how to get around unreasonable regulations and greedy officials. The two men would in a sense be joint expedition leaders, with Joe in charge of the scientists and Jiro responsible for the Burmese support staff.

The night before the flight north, Daw Marlar hosted the scientists at a farewell dinner and show at a concrete palace on a wide, misty lake near the Shwedagon. The Americans arrived at a quarter past seven, according to Joe's watch, fifteen minutes after the appointed time. Daw Marlar testily chided them, "You are late. Half hour late." The food was a bland, touristy version of *mohinga*, the Burmese national dish of freshwater fish and rice noodles; the show was a bland, touristy medley of classical music and dance, and some lame comedy sketches. Nobody laughed. Their thoughts were on the great enterprise that would begin the next morning.

Jiro was late meeting them at the hotel, so Joe took advantage of the unexpected sliver of downtime to post off a final batch of e-mails. He sent Sandy a short message of farewell:

> *Hi Sandy,*
>
> *I am "fixing to leave" (my Louisiana heritage) in a few minutes here for the north. Everything has gone fine and we are finally starting the fieldwork. We have something like 40 bags for 12 people, so it should be interesting at the airport this morning. I'll e-mail when I get back. I have no plans for that weekend of Oct. 12 so let's see each other on Friday.*
>
> *Miss ya.*
> *Cheers,*
> *Joe*

The flight north was no scarier than any other on Myanmar Airways: The airline's domestic fleet consisted of one ancient Fokker 10 that was always breaking down—and hard to repair, since international sanctions made it difficult, if not impossible, to get spare parts. The copilot's window was covered with yellowed newspaper as a sunscreen. The descent into Myitkyina, the capital of Kachin State, was a rough one as the plane skittered down through the remnants of a thunderstorm. In Myitkyina, the passengers had to wait while the pilot ate his lunch. Joe and a few other team members sat at the only empty table in the airport lounge to have a cup of tea, not realizing it was the VIP table until the waiter chased them off.

The flight to Putao was short, scarcely half an hour. In his journal, Joe described his first glimpse of the town: "As we descend, it is quite beautiful—verdant and wet—mountains in the distance." The storm they had crossed in Myitkyina had just blown through: A fresh breeze swept the sky with the brisk, clean smell of ozone, mitigating the heat.

The airport at Putao was a ramshackle shed with broken windows in the middle of a broad, stunning plateau encircled by snowcapped mountains. People in Putao say the mountains are shy: Cool and aloof, they keep their distance.

Moving from one state to another in the loosely confederated Union of Burma was tantamount to international travel; it took half an hour for the scientists to pass immigration control. Jiro delivered their passports in a stack to some young soldiers, who laboriously copied out the information by hand in ledgers; meanwhile, the group's bulky luggage was being hauled in from the plane on wobbly wagons drawn by stooped old men. In his message to Sandy, Joe had drastically underestimated the luggage count: The expedition checked sixty-one bags in Yangon.

U Thin Aung, the director of Hkakabo Razi National Park, was there to offer an official welcome to the scientists—an interesting development, since they were officially forbidden to go there. Also present was the manager of the advance team from Thurein Travel, who had been charged with building the jungle camps for the expedition and organizing the porters. His name was Nyunt Khin (confusingly, an inversion of the name of the company's powerful founder, Khin Nyunt). He would serve as the expedition's cook. While they were waiting at the airport, Htun Win pointed out to Joe that all the people working for Thurein appeared to have been hired from other travel companies. They were doing none of the work themselves. That didn't seem significant to Joe; after all, Thurein was a brand-new company.

As they waited in the luggage shed, Bruce Bartholomew noticed that Pocho Soe Lwin was hugging himself and shivering feverishly on a bench. Joe went to have a look at him, and sure enough, he had come down with malaria. Joe was annoyed: Before they left Yangon, Lwin had told him that he felt fine. Joe told Htun Win to put him on leave and get him to a doctor; he could rejoin the expedition after he began to recover. It seemed a bad omen to have someone fall sick before they had even made it out of the airport. Malaria was pandemic here; Joe knew that Lwin would probably not be the only member of the expedition to come down with it.

The naturalists stayed at the guesthouse attached to the army base on the edge of town. The guesthouse was clammy and densely populated by mosquitoes. The huge, sparely furnished rooms were painted a mossy shade of green, which took on an otherworldly glow in the garish fluorescent light provided during the two hours of electricity in the evening. Christiaan Klieger and Mark Moffett promptly dubbed the place the Emerald City. Many expedition members aired out their rooms to get rid of the dank reek of mildew and, as a result, spent the night slapping at mosquitoes.

As usual, Joe roomed with Dong Lin. That tradition had come about because of Joe's funky feet. He was embarrassed about how bad they smelled, and had never wanted a roommate; but Lin didn't care. On their third trip together to Burma, however, the odor had been so overpowering that Lin put Joe's sneakers in a Ziploc bag and hid them in the closet.

Putao was a one-stoplight town: At the main intersection, mounted on a pole, was a crude painting of a stoplight, with STOP, LOOK, and GO written in Burmese on the red, yellow, and green circles. The four great institutions of Putao were located at that intersection: the market, the infirmary, the Union Solidarity and Development Association (town hall, which showed pirated martial-arts videos in the evening), and the Kham Su Ko restaurant, the only decent place for visitors to eat.

There were no cars in Putao to obey the painted stoplight's commands: The town's transport fleet consisted of ten jeeps, perhaps the same number of decrepit trucks (several of them held together with baling wire), and almost as many Chinese two-stroke motorbikes as there were adolescent boys to drive them up and down the town's two main streets.

The town's sole cultural institution was a small museum that the Wildlife Conservation Society had created at the park headquarters, a handsome three-room frame house on the outskirts of town. The collection consisted of animal skulls and skeletons, stuffed birds, and enlarged photographs of Alan Rabinowitz in the field.

The region around Putao is home to hill tribes that migrated there from China more than four hundred years ago. These mountain peoples follow customs and beliefs different from those of the Burman, or Bamar, people, the nation's ruling majority, who control the fertile flatlands of the south. Known collectively to outsiders as the Kachin (a corruption of Chinese *ye-jen*, a derogatory term meaning "wild men"), they call themselves by the names of their own tribal groups, principally the Jingpaw, Rawang, and Lisu. Like other minority peoples in Burma, such as the Karen, Naga, Mon, Shan, and Arakanese, who occupy the country's hilly margins and the coastline, the Tibeto-Burman–speaking peoples known as the Kachin have never fully submitted to rule by the central government. When the British finally gained control of Upper Burma in 1885, they left the territory largely unadministered. After independence, the military government in Rangoon waged a futile war to suppress the Kachin, who resisted fiercely. In 1994, the Kachin Independence Army and other resistance groups reached a ceasefire accord with the central government; nonetheless, in 2001, most of Kachin State was beyond the army's direct control, and Burmese in name only.

At the Kham Su Ko restaurant, efficiently managed by a tall, grave Kachin woman and her sulky, long-haired son, the scientists had their last dinner by electric light. Nourished by greasy Chinese food and lubricated by beer, the group unwound: Finally there was the buoyant sense that an adventure was beginning. Joe formally introduced Jiro to the group, saying, "This is Jiro: He's our hero. We're going to give him a big tip at the end of the expedition—*if* he gets us where we want to go."

After dinner, they moved on to rum. Joe and Doug Long talked excitedly about the skull of a takin (*Budorcas taxicolor*, an alpine bovid that resembles a small musk ox) they had seen earlier in the day at the Hkakabo Razi National Park headquarters. "We're takin heads," quipped Long. In his account of an expedition to Putao District in 1921,

In Farthest Burma, Frank Kingdon-Ward wrote of the takin, "Much remains to be discovered, especially as regards the distribution of this animal, half-goat, half-buffalo." Eighty years later, the situation was unchanged. The takin skull Joe and Doug Long had seen at the park headquarters had some fur on it, which was unusual. The fur struck Long as being much darker than usual for the species, almost as if it had been smoked—certainly possible for a trophy kept in a Kachin house, which has the cooking fire in the middle of the main room. Joe suggested that they go back to the park headquarters and filch some hairs from the skull for DNA sequencing, to determine whether the skull was representative of a new subspecies of Budorcas, or possibly even a new species.

Problems were glimmering into view. Earlier in the day, Htun Win had confided to Joe that he was paying for many incidental costs out of his own pocket: some overweight baggage charges in Yangon, tips for porters, beers and soft drinks at lunch. Joe was furious. He scribbled in his journal, "What am I paying $44,000 for? Why do we have to pay for everything?" When he confronted Jiro about this, the guide explained that the $44,000 did not include drinks—not even water. Joe was shocked. Jiro promised to pilfer from the money Daw Marlar had given him, to pay for some drinks. Joe told him, "Just be sure you bring enough rum."

In the morning, Joe was up at first light, a gray paleness dripping with gentle rain. He and Doug Long walked through the mist into town with Jiro, about half a mile, to have a look at the market. Judged by its merchandise, Putao was a poor place. At the central market of a town this size in neighboring Thailand or China, you would find a much larger selection of foodstuffs and household staples. In Putao, the produce was dwarfish, the meat unwholesome-looking and expensive, and dry goods were limited to a small selection of shoddy Chinese products.

Doug Long was looking for animals and animal parts: Some of the most significant discoveries of modern zoology had been made in rural markets such as this one. In Putao he didn't find much apart from a tiny live monkey for sale as a pet, and a dead squirrel, which he bought. Joe bought a big bamboo drinking cup for the trip. They ate breakfast with Jiro in the market, sticky rice noodles under the tarp of a food stall. Rain was falling steadily.

When he rejoined the group, Joe found himself embroiled in the genteel, scholarly equivalent of a rebellion. Although he was the leader, the command structure of this expedition was more communal than was normally the case. In most field expeditions there is a P.I., the principal investigator, who raises all or most of the money and serves as a boss of sorts. Yet this group had cobbled together the money to pay Daw Marlar from several different sources: Bruce Bartholomew, David Boufford, Mark Moffett, and Christiaan Klieger came on grant money of their own; Joe was responsible for Guin Wogan, Doug Long, and Moe Flannery, as well as the Burmese field team, with funds coming out of a grant from the National Science Foundation; Dong Lin's expenses were paid by Cal Academy. To a greater degree than usual, there was the potential for factionalism or even mutiny; some of the people barely tolerated each other, and Joe was in no position to tell most of them what to do.

It was a bad sign that the stress lines were revealing themselves even before the group set out. The main cause for discontent was the inadequacy of the preparations for the expedition, which was becoming increasingly obvious. The botanists were annoyed because not nearly enough kerosene had been procured to power the dryers they would use to preserve their specimens, and there was no more for sale in the market in Putao. There were no tarpaulins to cover the scientists' bags, although conditions were undoubtedly going to be very wet. Bruce Bartholomew, the Cal Academy botanist, was particularly exasperated: He had written Joe just before they left the States to remind him about the need for kerosene and tarps. A cursory inspection revealed that the supplies of drinking water and rice were also scant; assurances from Nyunt Khin and Jiro that stores had been laid in along the trail fell on skeptical ears.

In Putao, they were at the end of the supply line, eight hundred miles north of Yangon; they were ready to move out, but they didn't have what they needed. Joe was just as annoyed as the others: Tarpaulins were distinctly enumerated among the supplies that Thurein Travel had promised to provide. Where were they? Jiro pleaded innocent and moved quickly to fill the gaps. There were no tarps in the market, but he found some long plastic tubes that could be adapted for the purpose. He wangled some kerosene out of the army, and the rest was procured at the airport from the wing of the airplane that had brought them in.

The most critical shortfall wasn't so easily solved: the trained medical professionals that Daw Marlar had promised. When Joe asked Jiro and Nyunt Khin who and where they were, all he got were shrugs and excuses. As usual, Jiro claimed that it was the first he had heard about it. Joe called Daw Marlar—not so easy, as there was only one telephone in Putao with a link to Yangon, which was located in a rustic radio shack on the outskirts of town. Her answer was that Jiro and Nyunt Khin were experienced in first aid. "Don't worry," she said, but the words had lost their soothing magic.

Joe challenged Jiro, asking to see the first-aid kit; it contained only a small assortment of basic supplies, with nothing in the way of medicine except aspirin and a few other over-the-counter remedies—no antibiotics, no antihistamines, no painkillers. Jiro mentioned that it might be possible, in the case of an emergency, to get a military helicopter to come on a rescue mission. Joe didn't have to point out how difficult that would be without a two-way radio, which they, as foreigners, weren't allowed to bring into the field.

It was Christiaan Klieger who tentatively suggested at this point that the group ought to reassess whether the expedition should go forward: In his view, the lack of basic emergency medical care raised the bar on risk too high. Dong Lin asked Joe about that: Didn't the Academy have guidelines to follow in such circumstances?

"Fuck the Academy," replied Joe.

"OK, brother," said Lin. "You're the chief. You make the decisions."

As unhappy as Joe was about the way things were shaping up, he never seriously considered a retreat: Too many busy scientists had gone to too much trouble to turn back empty-handed. Anyway, Joe had never been a great believer in taking elaborate precautions. His idea of emergency planning was to stay lucky.

If the other scientists were angry about the preparations he had made, nobody wanted to confront him publicly—his friends out of affection, not wishing to make him feel worse than he did, and the others out of fear. Joe was a genial and easygoing leader, and by nature democratic: He looked out for the porters with almost the same solicitude as he did the scientists. But he was also notorious for his intolerance of whiners. "There's no reason to complain," he would say. "We're all in the same situation." His answer to complaints, the polite one, was: "No cranky." Joe would do anything he could to help you, but he had no time for useless griping.

It took a few hours for the expedition to get under way. A lot had changed in biology since the days of Frank Kingdon-Ward, but the methods of field expeditions in Burma were about the same. The number of porters Jiro hired eventually came to one hundred and thirty. They were poor country people, mostly young men and adolescent boys clad in plaid *longyi*, sandals, and little more; there were a few young women, too. The porters were divided into syndicates, each one comprising about ten people from the same village. They would carry the scientific apparatus, personal gear, camping equipment, and food and drink for the expedition. The scientists were sixteen: ten Americans (eight men and two women), four Burmese, and two Chinese. There were eight mules for the heavy gear: the botanists' specimen-dryers, two generators, and the kerosene to power them. Except for the heavy machinery and the substitution of shorts and T-shirts for khaki, the group could have passed for a team of doughty Victorian naturalists.

The mustering point was Machanbaw, a minor British outpost in the days of the Raj, now a poor little town on the edge of the forest. It was the end of the road: Trucks ferried the supplies and the scientists there from Putao. The porters arrived in small groups, reporting from villages throughout the district. They all converged at an elegant suspension bridge, built in 1998, which spanned the eight hundred feet across the Mali Kha River, a raging torrent after a night of heavy rain. The scientists lined up on the bank of the river for a group photograph.

Joe, wearing the expedition T-shirt, acid green with the image of a striking cobra, stands in the center, surveying the scene with proprietary ease. On his left is his Burmese protégé, Htun Win, slim and looking even younger than his twenty-five years; on Joe's other side the two American botanists, Bruce Bartholomew and Dave Boufford, stand in a quiet group with the Chinese scientists and Aung Khwi Sein, the other Burmese herpetologist. The women stand together: Moe Flannery's pale face shines with excitement; Guin Wogan turns toward Joe with a confident smile. Ichthyologist Dave Catania beams steadily at the camera; Doug Long in a Hawaiian shirt, hands planted jauntily on his hips, looks as though he has just tossed off a joke. Christiaan Klieger clutches a ceremonial Rawang hat adorned with boar tusks, next to his friend Dong Lin. Mark Moffett, arms akimbo, stands by himself, at the end. (Hla Tun took the photograph.)

That night, the group stayed at an old British guest lodge, one big room for everybody. There was a sit-down toilet, a symbol of Machanbaw's former connection with the great world beyond Upper Burma, but it hadn't functioned in many years. The porters, as usual, had to shift for themselves—sleeping rough, or on someone's porch if they were lucky.

In the evening, Joe and a few others accompanied Christiaan Klieger to attend a service at the town's Baptist church. The road from Putao to Machanbaw was lined with Christian churches, mostly Church of Christ, Assembly of God, and Baptist, rude one-room structures roofed with rusting corrugated tin, and simple wooden crosses surmounting their bamboo entrance gates. According to the government in Yangon, 90 percent of Burma's forty-seven million people are Buddhist; 4 percent are Christian, and many of them inhabit the mountain fastnesses of Kachin State. The relics of Buddhism are everywhere around Putao, such as a giant pagoda bell made from the propeller of a wrecked

World War II fighter plane, but since the mid-twentieth century a substantial majority of the people there have converted to Christianity.

The first Christian missionary in the land of the Kachin was Eugenio Kincaid, a Baptist from Westfield, Connecticut, who poled and paddled his way there from Mandalay in 1837, in a boat full of Bibles and religious tracts. He described the Kachin as a "vastly interesting and affecting people" (considerably more generous than a British chief commissioner named Albert Fytche, who characterized them a few years later as "dirty, ugly, and barbarian"). In 1878, a missionary named William Henry Roberts, who had served under Robert E. Lee in the Army of Northern Virginia, petitioned Thibaw, the last king of Burma, for the right to build a church to minister to the Kachin. The king was a devout Buddhist and had no use for Christians: He contemptuously granted Roberts land "the size of a buffalo skin" and told him he might as well try to teach dogs to read. Undaunted, Roberts bought a buffalo skin and cut it into narrow strips, which he tied end to end and used to measure out the perimeter of a suitable compound for his ministry.

Perhaps the most intrepid evangelizers in Burma were Russell and Gertrude Morse, Church of Christ missionaries who had converted thousands of people in Putao District and established some seventy churches by 1965, when the government ordered them out of the country. One of their daughters was eight months pregnant and therefore could not fly on the military plane offered to the family for their removal. Rather than abandon her, the Morses decided to trek the seventy miles to the Indian border. They led their brood of children and a host of Lisu followers on an epic wilderness trek to rival that of Moses and the Israelites: climbing up and down sheer cliffs, building bamboo rafts to cross raging rivers, and feeding on fish, worms, beetles, nuts, fruit, and the rotting pith of tree ferns.

The Morses and their followers settled in a secluded valley about four miles shy of the Indian border, where they equipped their new mission with the ingenuity of the Swiss Family Robinson. The school's blackboard was made from a slab of chinquapin wood painted with a mixture of egg white and carbon; rolled-up cylinders of clay served as chalk. The Morse boys learned how to make gunpowder from charcoal, saltpeter, and sulfur, and used it to hunt monkeys and gibbons. The resourceful family melted down scrap aluminum from American aircraft that had crashed in the jungle during World War II, and cast their own pots and pans and spoons. The Morses' idyllic life in remote northern Burma continued until some of the children had to leave to get medical care for chronic ailments such as beriberi, brought on by malnutrition.

After the church service, Joe and Jiro had a long-postponed meeting to discuss logistics. Jiro told him the group could spend several days at each camp, but he needed advance notice. Joe explained that this was impossible. None of the scientists had been to the area before, so there was no way to predict which places would justify longer stays. Joe unfolded a map so they could plot an approximate route and itinerary, but every time he tried to pin the guide down about the exact location of the famous jungle camps that Daw Marlar's advance team had built, Jiro would slither around the subject, first talking about one route, then quickly changing to another. At this meeting, Joe concluded that the camps did not exist, although it seemed that some food supplies actually had been stashed along the way, as promised.

Joe proposed a plan that called for the team to stay two nights at some villages, allowing for a full day of fieldwork, and only one night at other camps. Once they made it to Pangnamdim, the entrance to Hkakabo Razi National Park, the scientists would have plenty of time to explore the subalpine terrain in depth. Based upon Jiro's descriptions of the trail, Joe made a tentative schedule for the first four days; after that they would create a new plan. Jiro went along with this, but pointed out that the advance team had only left enough food for one night at each camp.

The problem wasn't simply Jiro's inability to give them a straight answer, or Daw Marlar's failure to make good on her commitments: One of the biggest obstacles to drawing up an itinerary was the essential incompatibility among the scientists. The main fault line lay between the herpetologists and the botanists, and the sticking point was

DONG LIN

Joe Slowinski with the Burmese spitting cobra, *Naja mandalayensis*.

DONG LIN

Joe Slowinski in Burma. ∞

HLA TUN

DONG LIN

DONG LIN

MARTHA CROW

The scientist: *(top)* in the herpetology lab at the California Academy of Sciences; *(middle)* the opening of the Venoms exhibition at the museum in 2000; *(bottom)* Joe Slowinski and Dong Lin.

DONG LIN

BRIAN ARMSTRONG

BRIAN ARMSTRONG

NANCY DONNELLY

On the road with National Geographic: *(top)* Joe and Brady Barr examining a lively many-banded krait (*Bungarus multicinctus*), near Pakokku, Burma; *(middle)* Capturing a cobra in the road; *(bottom)* Lecturing at a training course for Burmese forest rangers. ALL PHOTOS COURTESY OF NATIONAL GEOGRAPHIC TELEVISION.

DONG LIN

ASHLEIGH B. SMYTHE

HLA TUN

HLA TUN

DONG LIN

(Top left) Joe in Peru; *(top right)* with stuffed snake; *(middle)* dancing after class in Burma; *(bottom left)* arm-wrestling in Burma; *(bottom right)* sampling an insect. ∞

HLA TUN

The expedition prepares to set out. September 1, 2001, at the bridge over the Mali Kha River, near Machanbaw, Burma *(from left to right):* Rao Dingqi, U Thein Aung ("Jiro"), unidentified Burmese Forest Ministry official, Aung Khwi Sein, David Boufford, Dao Zhiling, Bruce Bartholomew, Joe Slowinski, Htun Win, Maureen Flannery, Guinevere Wogan, David Catania, Douglas Long, P. Christiaan Klieger, Dong Lin, Mark Moffett.

DONG LIN

On the trail to Rat Baw.

DONG LIN

JAMIE JAMES

The bridge to Rat Baw.

JAMIE JAMES

Rat Baw schoolhouse.

DONG LIN

Joe Slowinski at the Moyingi Wetland Wildlife Sanctuary, Burma, in 1999.

pace. Botanists move slowly: Lord Cawdor, a Scottish earl who once accompanied Kingdon-Ward on an expedition to the Tsangpo gorges in Tibet, wrote of the experience in his diary, "In the whole of my life I've never seen such a slow mover—if ever I travel again I'll make damned sure it's not with a botanist." Anthropologist Christiaan Klieger, who needed to gain the trust of the people he met, allied with the botanists; indeed, he would have been quite happy to spend a couple of weeks at every stop along the way.

Herpetologists, on the other hand, tend to move quickly, in order to cover a lot of territory and maximize their chances of discovering animals—and Joe Slowinski had a reputation even among his fellow herpers for moving *fast*. Some of his best friends simply refused to go into the field with him, because he would impatiently abandon a site and move on if he didn't find interesting herpetofauna within a few hours. Mammalogist Doug Long sided with Joe; he, too, preferred to cover as much territory as possible. Joe referred to botanists as "salad scientists" with mingled affection and contempt, though in what proportion it would be difficult to say. The split between the botanists and the herpetologists was predictable, arising from fundamental causes. However much they might have in common as scientists, people who pick ferns and people who chase cobras have fundamentally different temperaments.

After the inconclusive meeting with Jiro, Joe cleared his head with a quick hike through the village environs, scouting for a place to go collecting in the evening. Herpetologists, like most of the reptiles and amphibians they seek, are active by night. It was raining lightly. There was no forest in the immediate vicinity of Machanbaw; the land was developed for agriculture, mostly rice, corn, and an ubiquitous leaf vegetable resembling the mustard green, which would become a staple of the expedition's diet. Joe's practiced eye saw at once that the area was unlikely to yield any unusual fauna. He found a muddy brook running through the village, which would have to suffice for the expedition's first night of collecting. The local people believed firmly in the existence of a giant water serpent they called the *bu-rin*, forty to fifty feet long, which attacked swimmers in such waterways. Tales like that were of more interest to Christiaan Klieger, the anthropologist, than to Joe. No one, it seemed, had ever actually seen the monster.

At dusk, around half past six, Joe set off with Wogan and Flannery, Burmese herpetologist Aung Khwi Sein, and the photographers, Hla Tun and Dong Lin. As the scientists descended the bank, Hla Tun fell and rolled into the water, an indignity Lin captured on video. Soon it began to pour rain, so Joe and Guin Wogan simply hopped into the brook. Christiaan Klieger, looking on, thought they were crazy. From behind they heard a cry of disgust: It was Lin, who had a two-inch-long leech attached to his leg. He leaned over with his camera to film it briefly. The outing was a washout, in every sense.

When they turned back, Joe caught a Sinoratrix, a common water snake, swimming between Moe Flannery's legs; Wogan had collected a few frogs. At the lodge, to take the damp out, they drank potent army rum, distilled by the soldiers at Putao, who infused it with bitter quinine for its antimalarial properties.

It was an early night. Joe ordered breakfast at five thirty, for a six o'clock start.

"Of course, breakfast is not ready at 5:30," Joe's journal began the next day. It took hours to load up the porters; then, when it became apparent that there weren't enough shoulders to carry all the gear, Jiro sent to some villages nearby for more men—yet another delay. The herpers used the hiatus profitably, to fix the few specimens they had caught the day before. Finally the unwieldy group, nearly a hundred and fifty strong, jerked into motion. It was twelve miserable miles to Alanga.

It rained all day, hard and steady, with never a hope of clearing. The trail was a mockery of a road: It looked like a road, with deep ruts from the trucks that came along occasionally in the dry season, but in the middle of the monsoon it was a river of mud, an infinitely fine clay that possessed miraculous powers of water absorption and therefore of suction—strong enough to pull a boot off.

The trail was thronged with leeches. There were legions of the bloodsuckers on the ground, marching along in their grotesque, leaping way; they clung in thirsty clumps to the brush alongside the road and hung from the foliage of the trees like tiny vampire bats. In *Burma's Icy Mountains*, Kingdon-Ward described the predatory behavior of the leeches on his expedition to Putao District in 1937: "It was rather horrible to see the hordes of famished leeches advancing immediately one entered the jungle. It is almost indecent how they smell their victim and sway their way towards him, the foliage shivering to their regular movements."

In addition to the ordinary leeches, an inch or two in length, the hikers were also plagued by big green leeches, which looked like leaves. These parasites swam through waters bloodied by the other leeches and affixed themselves to the hikers' feet. It gave the naturalists great incentive to keep moving; to pause was to get bitten. Dave Catania had blood streaming from his beard, from bites by leeches that had dropped from tree branches to burrow in it. Doug Long sported women's pantyhose, made of synthetics so densely woven that (in theory, at least) the leeches' suckers couldn't pierce it. Mark Moffett tried a more manly repellent, spitting tobacco juice on his legs to ward off the parasites; Christiaan Klieger created a hybrid, wearing socks stuffed with tobacco leaves over his pantyhose. Despite the elaborate precautions, everyone got bitten, sometimes by many leeches simultaneously: They would form rows, like suckling puppies.

The extreme, arduous conditions tested the team's mettle to the limit. Most of them had experienced terrible ordeals on the trail, but nothing like this; tempers were getting seriously frayed. Moe Flannery, on her first expedition, seemed cheerfully oblivious to her mud-caked condition, tramping around camp like Calamity Jane. This seemed to annoy Bruce Bartholomew, who struggled to maintain a natty appearance despite the squalid conditions. Jiro was getting on people's nerves with his hollow promises and incessant excuse-making.

Joe followed his usual practice: He ignored the discomfort. He wore shorts and sandals, and simply let himself get wet and muddy and bitten. To distract himself from the awful conditions, he carried his Discman, blasting a CD from his weirdly assorted collection: Chuck Berry or Judas Priest, the Beach Boys or the Sex Pistols, and ABBA, always ABBA. Joe had long ago adopted the gay disco anthem "Dancing Queen" as his theme song, getting a lift from it as he rambled along forest trails throughout the world. When straight-laced colleagues, more inclined toward Mozart or John Denver—or silence—teased him about it, he would shrug cheerfully and say, "I don't care, I like it. It's a good song."

An hour into the march, Joe caught a nice *Ahaetulla prasina*, the common Oriental whip snake. Before leaving for Burma, Joe had posted some beautiful photographs of this vine snake on his Web site. The photos, taken by Dong Lin, showed the snake passing through several color phases, from bright green to corn yellow to bluish white. *Ahaetulla prasina* has a pointed, Concorde-like snout that enhances its binocular vision and, therefore, its hunting prowess.

That same day, Joe had one of the most reckless exploits of a career already full of Pecos Bill–style heroics, which he described in his journal:

"Soon we come to a suspension bridge over a large river. Much to my dismay, an adolescent bull marches onto the bridge. Then it charges me. I butt it football-style, which knocks it on its ass. It skids across the wet bridge and nearly falls off. Then it runs across the bridge to attack the porters." Joe had the porters as witnesses, otherwise the incident almost defies credulity: The skull of a bullock is as hard as an oak stump. Joe knew no fear of animals; it often seemed that he thought of himself as one of them. He used to tell Dong Lin, "I'm a wild animal."

It was nearly dusk by the time the expedition reached Alanga village, the camp for the night. It seemed that the advance team had indeed been there—not to build a camp for the scientists, who would stay in the thatch-roofed schoolhouse, but to construct temporary shelters for the porters, flimsy bamboo huts that were little more than lean-tos.

Alanga was a wretchedly poor place. Every village in Kachin State is poor, but there are many shades of want: Poverty is to Burma what snow is to Eskimos. A few prosperous folk in Machanbaw possessed such luxuries as tin roofs and battery-powered radios (though not television—there are no television broadcasts in Putao District), but Alanga was a place with few resources beyond the hardiness of the people. The houses, rickety bamboo structures on stilts with thatched roofs, usually had two rooms, a common living area adjoining a smaller bedroom. The livestock, if there were any, huddled with their attendant vermin on the ground under the house. The cooking fire was in the middle of the larger room, with no ventilation other than the gaps in the floor and walls. Most Kachin houses have no furniture except a bed and perhaps a teak chest to protect cloth from the ravages of rot and insects.

Malaria is rampant; there is virtually no medical care except herbal cures and magic; education, where it exists at all, is rudimentary; and no one eats a decent diet. Yet villages such as Alanga are pleasant places, the people polite and hospitable. Here, as in every hamlet the scientists passed through, the expedition's arrival created a stir. It was the most spectacular event of the year. Joe was friendly to everyone who was brave enough to make eye contact, and he gave away lots of little gifts—ballpoint pens to the children, bars of soap for their mothers. Four years later, people in Alanga and the other villages on Joe Slowinski's route remembered the Cal Academy expedition well. Everybody remembered Joe; his picture always ignited a bright smile.

As soon as the naturalists arrived at the schoolhouse and collapsed from exhaustion, Joe told Jiro to get the word out to the people in the village that he wanted to buy frogs, lizards, and snakes, dead or alive. The going rate was two hundred kyats (about twenty-five cents) for nonvenomous snakes, two thousand for a cobra or viper. Mark Moffett was in the market for insects. Nearly a hundred people came in, peddling all sorts of creatures: beetles, praying mantises, frogs, but only one snake. Doug Long bought the skull and fur of a goral, an endangered species of antelope, for five thousand kyats, about six dollars—big money in Alanga.

After dinner, Hla Tun, the Burmese team's photographer, told Joe that Htun Win was very unhappy with the way things were going. Joe knew that the talented young herpetologist was proud; now he was showing a rebellious streak. Later, Joe noted in his journal that the Burmese field-team leader was chafing at his subordinate position.

Joe slept in the next morning till eleven; the night before had been a late one, soused with rum. By the time he hit the trail, most of the others had been on the road for hours, but moving at his usual headlong pace, he eventually caught up. The march to the next camp, a village called Htanga, was even harder than that of the day before. The grade was steeper, the mud was deeper, and the leeches even more abundant. Most of the scientists wore Chinese military-issue boots, which afforded some protection against the insatiable, bloodsucking worms—at least until they wriggled their way inside the boot. The porters had no protection at all for their feet except calluses; in their wake, the puddles turned a sickening shade of crimson, fading to muddy pink as the rain sheeted down.

Jiro had told Joe it was three hours to Htanga—a gross underestimate for a grueling fourteen-mile hike. He led the expedition through the forest on a *shat khat*, good colonial-era Burmese for "short cut," which was even muddier and more difficult to negotiate than the main road; but the path led through the first pristine wilderness the scientists had seen. They heard the screech of gibbons in the treetops. Doug Long was amazed at the number of rare and unusual birds he saw; Mark Moffett reported seeing a pair of majestic hornbills flap by overhead. These were the heralds of intact wilderness ahead.

The road climbed to an elevation of nearly two thousand feet, through a rugged jungle of towering dipterocarps, Chinese coffin trees, flowering magnolias, fragrant screw pines, and many fruit trees, including rambutan, mangosteen, and banana. Kingdon-Ward described this terrain after his final expedition there, in 1953: "Here the forest is richer and denser—not only does frost never enter to these deep sheltered valleys, but throughout the winter they are steeped in mist till nearly midday, and so partake of the character of tropical rain forest."

Trees were wrapped with lianas and other climbers. In the shady recesses of the jungle there were primeval-looking fern trees and lovely orchids, including the rare black orchid.

It was nearly sundown by the time Htanga came into view: a good-size Lisu village of perhaps forty houses ranging across hillsides overlooking a bend in the narrow river, made torrential by the relentless rain. The expedition entered the village like a weary army in retreat, crossing a graceful cable-suspension bridge in small groups, a few at a time. In the humid twilight, the wet thatched roofs and bamboo fences of Htanga were almost indistinguishable from the surrounding forest.

Htanga felt more remote from the modern world than any place the naturalists had yet seen. The houses there had small spirit shrines for the nats, deities midway between demigods and fairies, sometimes heroic and sometimes malevolent, which have been worshiped in Burma for thousands of years. When Buddhism was enforced as the state religion of the kingdom in the eleventh century, the nats were driven underground (where most of them lived to begin with); but soon they were syncretized into a greater Burmese cosmology, with the Buddha raised, in a rather un-Buddhalike way, to the first heaven. There are shrines to the nats at the Shwedagon. The cult of the nats remained strong until the end of the twentieth century; globalization exerts a feeble influence in Burma compared with the countries that neighbor it (or almost any place in the world), but many young people in the cities now neglect the nats. Nat shrines are much like spirit houses in Thailand: small bamboo cupboards where worshipers leave offerings of food and cigarettes to propitiate the fickle divinities.

As in Alanga, the schoolhouse in Htanga was made available to the scientists as a campsite, but it was much more squalid: The roof leaked, and the leeches had moved in and lurked in every damp, dirty corner. The scientists pitched their tents inside the school, cheek by jowl. After dark, sandflies swarmed. The tiny, bloodsucking insects left a nasty sting with their bite, which was much sharper than that of mosquitoes; they also sometimes carried leishmaniasis, an excruciating, potentially deadly hemorrhagic fever also known as Q fever. The generator had to be started up to recharge laptop and video batteries, and to run the black light that Mark Moffett set out at night to lure insects for his collection. Since it was raining, there was only one place to put the noisy, kerosene-reeking machine: under the floorboards of the schoolhouse.

It was a hellish night. When the generator was finally turned off, there was little relief, because the roar of nocturnal insects and frogs was deafening. The relentless buzz of the wet jungle bored like a dentist's drill into the nerves of even the most seasoned trekkers. Again Joe stayed up late, drinking rum from the big bamboo cup he had bought at the market in Putao.

And the next morning, once more, he slept in till eleven. There was no question of the expedition moving on to the next stage: They had walked twenty-six miles in two days, under horrendous conditions. At least a day of recovery was obviously needed. Also, as far as many of the scientists were concerned, the time to start doing some intensive exploring and collecting was long overdue. Dave Catania was out early, setting a fishing net in the river; Moe Flannery raised a mist net in a small clearing to catch birds.

Joe hurled himself into the jungle to look for snakes. He followed the river upstream from the village, hopping across granite and basalt boulders. It was pouring rain; leeches besieged him. Eventually he made his way to some ancient forest, but he came up nearly empty-handed. In his journal he wrote, "The herpin' around here sucks. We are finding virtually nothing." Joe's previous expeditions to Burma had been in the dry central plains, where cobras and vipers, the venomous snakes he loved, were plentiful. That day in Htanga, the best finds were insects and amphibians, and most of those were brought in by people from the village. There were some exceptional stag beetles, a handsome red-throated lizard, and a few unusual toads. By this point, despite the difficult conditions, the expedition had most likely already collected a few new species; they would determine that when they got back to the museum in San Francisco, where they could do some comparisons in the collection and research the literature.

Doug Long made a fascinating acquisition in Htanga: A villager brought in a tiny shrew with unusual dentition, which Long thought was likely to be a previously unidentified species. Shrews are among the smallest mammals on Earth; the Etruscan shrew, *Suncus etruscus*, which is widely distributed throughout Eurasia, is the tiniest of all. This tiny blob of life, measured by mass, sometimes weighs less than two grams. A few species of shrew are venomous, an extremely rare adaptation in mammals.

That evening, Mark Moffett did something very strange: He fed Long's shrew to a large mantis he was keeping. Long's ready sense of humor for once abandoned him; he was furious to have lost his specimen, one that might have been of great scientific value. Others in the group were both irritated and slightly unnerved.

All the scientists on the expedition were intently focused on their own research but pulled together more or less cohesively when the need arose; Moffett, however, was viewed as a loner, always off on his own—looking for stories for *National Geographic*, he said. And while many of the zoologists on the expedition shared Joe's fascination with the deadly power of nature, there was something bizarre—from a biological perspective, almost freakish—about feeding a mammal to an insect. It was tantamount to upending the food chain.

Joe had to take Moffett aside and give him a little talking-to, not a job he relished. He looked upon the unpredictable entomologist as a peer and a friend.

They had met two years before, when Joe and Dong Lin were attending a series of evening herpetology seminars in Berkeley. Before the seminar, the three men would meet at La Val's Pizza, a Berkeley landmark on Durant Avenue, just off Telegraph. Over a couple of rounds of Coronas, the herpetologist and the entomologist would swap their best stories, and Joe would talk up his next expedition, slapping Moffett on the back and saying, "Come along this time, bro." The Hkakabo Razi expedition was the first time their schedules had jibed.

In Burma, the two men had grown closer. Joe was a sociable night owl: When he came in from night-collecting, Moffett was sometimes the only one still up. The two would stay up late, drinking and talking. In an article about the expedition for *Outside* magazine, Moffett recalled: "Some nights it seemed he felt invincible. Downing Burmese rum, he knew he would rise high enough in the hierarchy of science to put a stop to the 'political bullshit' he saw all around him. . . . He confided a thousand ambitions, certain he'd realize them all." On other nights, Joe was melancholic and complained bitterly about Alan Rabinowitz's opposition to his research in Burma, sometimes working himself into a rage.

Sensing that his clients needed some homely comfort, Jiro supervised a nice dinner for them that night in Htanga: For the cocktail hour he opened a bag of imported potato chips to accompany the inevitable rum, and for dinner Nyunt Khin cooked up a tasty spread of fried pork, potatoes, and pumpkin. (That was one thing that had changed for the better since Kingdon-Ward's day; the plucky botanist was always battling rotten, weevily food. However, on his final expedition to Burma, in 1953, he celebrated Queen Elizabeth's coronation day in the jungle with a feast of pickled walnuts and maraschino cherries, roast cock, and a tinned Christmas pudding from Fortnum & Mason.)

After dinner, the herpetologists went out for a round of night-collecting. Htun Win volunteered to go on a solo walkabout: He would head into the forest with a porter and rejoin the group in Ureinga, the expedition's next encampment. The others divided into two groups: Joe took Aung Khwi Sein and Guin Wogan upriver, following the route he had taken in the afternoon, while Rao Dingqi, the guest herpetologist from Yunnan, led the others into the forest in the valley. It started raining almost immediately, but they soldiered on.

At last they found a few interesting herps. Joe loped ahead on his own, as usual, and collected a strange agamid, an iguanalike lizard capable of changing colors like a chameleon. The one Joe found sported a splendid scarlet dewlap. Wogan caught up with him and showed him a large, beautiful Megophrys, a leaf frog she had found sitting in the middle of the muddy trail. Shortly afterward, Joe found another of the same species himself. Rao Dingqi's group also had some luck, finding several different species of frog, including the rare Tenasserim frog

(*Rana livida*), which is known to exist only in Burma. After three hours, soaked through, the herpetologists returned to the schoolhouse, where the sandflies and mosquitoes whirled in tormenting clouds.

When the scientists woke up the next morning, a heavy rain was bucketing down. The river running through Htanga had risen by more than ten feet, raising the ominous possibility of floods. Joe had to make a decision: Should they press ahead today, as they had planned, or stay on in Htanga? He stuck a pole into the river to monitor its height; surprisingly, an hour later the water level had dropped five feet. While this positive trend tended to support the idea of pressing ahead, Joe was encouraged by the decent haul of specimens they had collected in Htanga the day before, and he decided to lay over for a third night.

The rain stopped in late morning. Joe hiked almost half the way back to Alanga, but found nothing. After a dinner of carp and bamboo shoots, the herpers went out for another night-collecting expedition. Joe noted in his journal that he found "basically nada," but Guin Wogan and Aung Khwi Sein captured a nice pair of Rhacophorus, tree frogs with orange spots on their backs, sitting on a branch overhanging a shallow pool of water.

When Joe returned to the camp, at half past ten, he found everyone zipped up in their tents to escape the maddening sandflies and malarial mosquitoes. They were either sleeping or, more likely, pretending to sleep, catatonic with exhaustion. Dong Lin had forgotten to leave a bottle of rum for Joe as he had promised. Joe shook him from his torpor and cadged a glass, and Moe Flannery gave him some precious whiskey from her flask: enough medicine to purchase another night's oblivion.

The morale of the expedition was beginning to collapse.

The next day the sun finally came out, getting even lethargic Joe up at an early hour. While Jiro struck camp and loaded the porters, Joe and Lin worked on the photographic archive, taking still pictures and videos of the frogs that had been collected the day before. The expedition set off at nine, an early start for this group. Their destination, Ureinga, was not a village at all but a camp for workers from the Kachin State Survey Department; they were supposed to be building a new road, though there was no evidence of any work apart from the camps themselves. The sunshine was welcome, but the ten-mile hike was no easier than the ones that had preceded it: The rocky road, with its thick coating of wet clay, was extremely treacherous to walk on. By the time the expedition reached Ureinga, around two in the afternoon, the seats of many pants had been muddied by painful pratfalls.

When the group finally arrived at the poor cluster of huts that would shelter them in Ureinga, they found Htun Win waiting for them. After the camp was set up, he showed Joe what he had collected. There was a good haul of frogs, including a fine Rhacophorus like the ones Wogan and Sein had found. Most intriguing, Htun Win had caught a Dendroaspis, an elapid closely related to the infamous black mamba of Africa. Joe was delighted: He had never seen one before. Then he had a strategic meeting with Jiro, which brought home to him how much farther they still had to go: "Yikes!" he wrote in his journal, "we won't have much time in Pangnamdim."

The sunny conditions did nothing to allay the predations of the leeches. When Joe went out to collect in the afternoon in shorts and sandals, he was assaulted. The sun also brought out dense swarms of bees and wasps in search of the salts in the sweat that dripped profusely from the naturalists' bodies in the humid heat. Every now and then, a slap and an exasperated curse would be heard from another victim. Doug Long sweated so heavily that he was attacked by a cloud of butterflies.

In Ureinga, the night-collecting proved to be good, better than in Htanga. Joe found a fine *Boiga quincunciata*, the Assamese cat snake, previously known in Burma from only a few specimens. The champion of the night was Aung Khwi Sein, who bagged a caecilian. Caecilians are burrowing, limbless amphibians that look like earthworms; because they spend all their lives underground, they are very difficult to find. Joe was sure that Sein's specimen set the northernmost record for a caecilian in Asia.

Everyone found something of interest in Ureinga: Christiaan Klieger did a long interview with the sole Kachin family living there full-time, with Jiro serving as a halting interpreter. Klieger was one of the most distinguished members of the group, with five books to his credit. He had in press an anthropological memoir called *Tibet-o-Rama*, about how hyper-romanticized, "orientalist" views of Asia in the West create confusion in cross-cultural romantic relationships—as it had in his case with a young Tibetan refugee. He was the first trained cultural anthropologist to work in northern Burma since Edmund Leach, a British student of Bronislaw Malinowski. In 1954, Leach published a classic study of the Kachin called *Political Systems of Highland Burma*, which challenged the accepted wisdom that the boundaries of society, culture, and language are necessarily congruent, and pointed out some errors in the work of his most illustrious predecessor in Kachin studies, Claude Lévi-Strauss.

On the following night, Joe went on a collecting expedition with Htun Win and Dong Lin, and returned with a few good frogs but no snakes; Aung Khwi Sein collected a fine specimen of the many-banded krait. Rao Dingqi, the Chinese herpetologist, caught a Parcas, a snake that feeds on snails. It has an efficient method for eating them: The snake holds the shell of the prey firmly between its jaws while a ratchetlike tongue darts into it and yanks out the living snail—the same engineering principle employed by the table implements for eating escargot.

The herpers did well by Mark Moffett, finding him a beautiful flower mantis with expanded graspers that concealed crimson patches. They also caught a two-foot-long walking stick, the ungainly insect that mimics a jumble of twigs as camouflage against predators. The scientists made a pet of the weird creature: At that night's rum-drinking session, they let it roam around the room in its comical, jerky way.

At Ureinga, the botanists were busy drying specimens; Jiro and a few of the more capable porters assisted them. The dried specimens were kept in cardboard accordion files, with aluminum covers to protect them. Jiro suggested that they cache the specimens they had collected up to this point in Ureinga, but Bruce Bartholomew and the other botanists were afraid they would spoil. Jiro finally convinced them that they would be safe there, so they left them behind.

In the morning, the group pressed ahead. At Ureinga, the trail veered north, aimed directly toward Hkakabo Razi. It was eleven miles to a Lisu village called Ba Baw. *Baw* means "ferry"; this was the place where travelers forded Ba Creek. The name had acquired an antique patina since a fine bamboo footbridge had been built over the creek, but there was still a bamboo-raft ferry in the dry season, for the occasional truck. The weather had changed; rain now was only intermittent. Yet this walk seemed especially punishing to Joe: His ankles throbbed with pain after many days of pounding his feet on wet rocks.

The scientists were weary and depressed. Many of them walked and worked in silence, to avoid quarreling. The flying and creeping bloodsuckers hadn't abated; that night at their camp in Ba Baw, a drafty bamboo house, the scientists put flaming logs under the dinner table to keep the sandflies away. Even eye- and throat-blistering smoke was preferable to the torture of the insects. Joe wrote in his journal, "My body has responded very poorly to the sandfly bites, which are very swollen." Most people would have complained about the pain inflicted by the insects, but with a scientist's objectivity, Joe simply observed their effect on his body. Guin Wogan gave him a Benadryl, which helped bring the welts down.

Those closest to Joe were beginning to worry about him. The extraordinarily difficult conditions of trail and campsite, the barely contained emotional strains among the expedition members, the disappointing haul of herps, the relentless cycle of extreme physical exertion alternating with immoderate alcohol consumption: all these things were taking their toll on him. Now when they walked on the trail together, Mark Moffett noticed that Joe moved sluggishly; when he paused to pluck a leech from his legs, his hand trembled.

The next morning, as usual, Joe and Dong Lin photographed the herps collected the previous day. Joe wrangled the many-banded krait so Lin could get in close, just inches away. Every child in the village came to gawk at the reckless foreigners; their parents hung back shyly, watching over the fence. It was a daredevil act Joe and Dong Lin had honed on many previous expeditions.

The land around Ba Baw had been developed for farmland, fields of corn and the piquant mustard green they ate every night at dinner. As a result, the chances of finding unusual wildlife were reduced, so a consensus quickly formed that they should move on in the morning to Rat Baw, their next stage. Jiro told them that in Rat Baw they would have a good campsite, a large, well-built schoolhouse, and some primary forest to explore.

On the road to Rat Baw, a porter killed a *Sinomicrurus macclellandi*, a find that held a special significance for Joe. He had just renamed this sleek, blood-red coral snake, formerly *Calliophis macclellandi*, in a paper coauthored with colleagues from Louisiana State University and Cal Academy and published two months before the expedition. Joe and his colleagues created a new genus, Sinomicrurus, to take into account certain genetic and morphological similarities among a group of Asian coral snakes that had previously been overlooked.

Joe hiked much of the way to Rat Baw with Dong Lin and unburdened his soul. He told his best friend that when he got home from this expedition, he wanted to settle down with Sandy. They had only known each other for a short time, but he felt sure: She was the one. What was more, he was sure that she felt the same way. He was traveling too much; this was already his third trip to Burma this year. He was exhausted, and he wanted a life beyond herps and herpetology. Lin was surprised, and a little upset. He said, "Hey, Joe, you mean you're not going to hang out with me on Saturday nights, drinking beer and playing pool?"

Joe replied, "That's right, bro."

Christiaan Klieger and Doug Long—always the slowest walker on the trail, to Joe's annoyance—were among the last in the group to reach Rat Baw. On a pretty stretch of road that followed the course of the river, just outside the village, the two men came upon a water buffalo calf standing on the side of the trail, shivering fearfully in the tall grass. It had somehow become separated from its mother. When the men came near the calf, it got confused, crashed through the brush on the riverside, and jumped into the river with a great splash. It was instantly swept away by the torrent, lowing piteously as it disappeared around a bend in the river, into the twilight gloom.

Rat Baw was the biggest town the explorers had seen since their trek began. It was tucked into a narrow valley between two high ridges topped with swirling clouds, spreading across both banks of the river. The settlement on the near bank, Rat Baw 2, had twenty-three houses; a long, high suspension bridge swaying over the roaring river led to Rat Baw 1, which had twenty-five households. The village had a charming, Tolkienesque air: Bamboo fences crisscrossed the gentle hillside, ruling off neat vegetable patches; the low roofs of the houses, thatched with fan-palm leaves, blended imperceptibly with the surrounding secondary forest.

A wide avenue with a cobblestone road, the first pavement the group had seen since Machanbaw, followed the river into the mountains, toward Naung Mung. Naung Mung was the last town of any size this side of Hkakabo Razi National Park; it was also the location of the closest army base. A few hundred yards down the cobble road, a dirt path curved back toward the river, leading to the schoolhouse—a solid, handsome frame building painted chocolate brown, with a newish tin roof. The school had five classrooms, fronted by a porch that ran the whole length of the building.

It was the most comfortable campsite of the expedition thus far. There was plenty of room for all the tents to be pitched indoors, and the school desks could be put to all sorts of good uses. The headmaster of the school, a dignified Jingpaw man named Joseph Tawng Wa, welcomed the foreign guests. He canceled classes for as long as the scientists wanted to stay there. Even the porters finally had dry places to stay; many of them bunked with the school staff in a long outbuilding running at a right angle to the schoolhouse, which created a small campus with a flagpole in the middle. The porters quickly built an impromptu kitchen nearby. The expedition would be snug in Rat Baw for a few days, if the collecting warranted it, which might give their war wounds a chance to heal.

The next morning brought bad news that overturned all their plans. It was exactly what Joe had been dreading since they left Putao. When he returned from his morning collecting excursion, the porters reported that there had been flooding and mudslides on the trail ahead. The bridge at Gawlei, the village midway between their present position and Pangnamdim, had been washed out. That effectively took Hkakabo Razi off the table. The river was much too dangerous to attempt ferrying across; the current was strong enough to wash away a car. And even if they did manage to cross at Gawlei, another bridge farther along was also washed out, the porters said. Joe, demoralized and depressed, saw the end of the road looming before him.

Yet there was one positive development: Rat Baw was a promising location for collecting herps. In the afternoon, after Jiro had circulated the usual invitation to the local folk to bring in animals to sell, the response was overwhelming: "Lots of stuff comes in," Joe wrote in his journal. "Birds, frogs, snakes, insects. The snakes are great." Most of them were colubrids, but there was also a good haul of elapids. A man in the village brought in a juvenile monocled cobra; Joe paid him three thousand kyats for it. The night before, Guin Wogan had collected another many-banded krait, a major addition to the collection. The accumulating mass of collection bags was finally beginning to look impressive.

The frog collection was already extensive, and now they had several turtles in two genera, Manouria and Cuora. Joe decided to euthanize and collect just two of them. It was a sensitive issue, as many Asian turtle species face extinction. After the severe criticisms of his collecting practices in Alan Rabinowitz's book, Joe more than ever wanted to err on the side of restraint. Dave Catania and Moe Flannery were pleased with their progress too. Flannery had caught a woodpecker that she was certain was a new species—a major accomplishment for her first field expedition.

That morning, while he and Dong Lin were photographing the new specimens, Joe made an interesting discovery: One of the snakes that had been identified as a krait was actually a Dinodon, the harmless colubrid that mimics the deadly venomous species so uncannily well. In the afternoon, the sky cleared. Moe Flannery played soccer with the schoolchildren, who were enjoying their unexpected holiday.

When Joe came back from his afternoon collecting rounds, he was leaping with excitement. He had found a pair of kraits mating in the jungle, entwined around each other like the strands of a lariat. "It was beautiful," he exulted to Mark Moffett, slicing his arms, semaphore-style, through the heavy air. "Goddamn *beau*-ti-ful! Courting like that, right in the middle of the trail. I've never seen anything like it." He bounced back to his tent to relax, popping *The Great Rock 'n' Roll Swindle* by the Sex Pistols into his portable CD player.

That afternoon brought another piece of good news. Hla Tun sought out Joe to tell him that a "telephone letter" from U Khin Maung Zaw, a message dictated over the army radio in Naung Mung, had arrived, informing the expedition that permission had been granted to bring six scientists from the Forest Ministry to the United States. It was a rare triumph of learning over bureaucracy in Burma. Zaw reiterated in the telephone letter that he wanted to be among the group going to San Francisco. When Joe told him the news, Htun Win was ecstatic; later, he visited Jiro in the kitchen while he was preparing the evening meal, to tell him how happy he was about going to America.

Finally, there was something to celebrate: A new woodpecker was about to find a perch in science, and the Burmese were going to California. Joe asked Jiro to organize a beef barbecue for the following day. Everybody was invited—the porters would eat beef too. That afternoon, Jiro bought a small cow and supervised its butchering. He sent porters into the village to buy *lokhu*, the local rice whiskey. When Jiro asked how much he should buy, Joe replied, "Get enough for a hundred and thirty porters." Jiro bought twelve gallons of the potent liquor, an enormous supply. Most Burmese people are averse to getting very drunk because they believe it invites possession by an evil nat; and in northernmost Burma, Kachin people are likely to be teetotaling Christians. It was understood that the *lokhu* was being laid in primarily for the foreigners.

Early in the evening Joe had a talk with Joseph Tawng Wa, the schoolmaster. He was the first person Joe had met on the trail who spoke real English; it was halting and accented, but articulate. They talked about religion. Wa said he was Catholic, and asked Joe about his faith. Joe told him he was a lapsed Catholic, a concept Wa understood and took in his stride. He was a grave, placid man with two gold incisors, who had lost three of his five children to malaria. He was happy to meet Joe, he said. Wa loved America. He carried a laminated postcard of Bill Clinton in his wallet.

That night, Htun Win was studying the herps that had been caught during the day. Pocho Soe Lwin had brought in two small snakes, which looked like Dinodons. Htun Win examined them with particular care, after Joe's discovery the previous day that a Dinodon had been misidentified as a many-banded krait. While Htun Win was handling the snakes, one of them bit him. By then he was sure it was a Dinodon, he said later. He quietly waited for a while, but experienced no toxic symptoms. Later that night he went into the jungle to collect on his own, without inviting Joe. That hurt Joe.

After dinner, the scientists sampled the *lokhu*. The Burmese stood in the shadows, watching the uninhibited foreigners. Thumbing his nose at the washed-out bridges, Joe was in manic high spirits. Flushed with Jingpaw moonshine, his face glowed scarlet in the harsh lantern light. At least they were collecting some interesting snakes now, damn it. He was sure there were a few new species among them, perhaps more than a few. In an agitated access of pride, Joe dragged the snakes, carefully sorted and labeled in their cloth collection bags, into the dirt road in front of the schoolhouse and piled them up.

Then he lifted the bags, all of them, and held them aloft in his arms. Dong Lin took the photograph. The expedition's chief porter, Yosep Kokae, a slim, shy young man from Putao, stood far outside the circle of lamplight, watching in wonderment as Joe crowed, "I am the king of snakes!" Whooping in the humid jungle darkness, Joe cried, "I can catch snakes better than anybody!"

The next morning, Joe was up at first light.

The schoolhouse at Rat Baw had five rooms in a row, all facing a wide railed porch that ran along the front of the building. Joe and Dong Lin shared the classroom at the southern end; the rest of the group pitched their tents in the longer room next to it. That morning, Joe came bounding in and gave Guin Wogan a shake. "Hey, get up," he cried. "Come on, let's go look at snakes!"

Wogan answered, "All right, all right, I'm there."

Joe shook the tent next to hers. "Come on, let's go!" He's just like a little kid, thought Wogan, always so impatient.

Joe clomped outside to have a look at the morning. The sky was still gray, but streaked with pale light, promising a fine day. Mist was already rising from the damp grass. Joe found Jiro in the kitchen shed behind the school, cooking fried rice with eggs and peanuts for breakfast. Joe took a cup of coffee and inquired about the preparations for the barbecue.

He wandered off to find Htun Win. As usual, the young Burmese herpetologist was already hard at work, sorting specimens. Guin Wogan was there, still looking sleepy, and so was Dong Lin. He and Joe both wore the expedition's bright green T-shirt. Mark Moffett emerged from the schoolhouse and walked up to join them.

Joe asked Htun Win about one of the specimens. Htun Win held the bag up and said, "I think it's a Dinodon."

Joe took the bag from him. Although it was still too dim to see well, he carelessly stuck his right hand into the bag. When he pulled it out, a thin, banded snake less than ten inches long was dangling down, its fangs sunk into the base of Joe's middle finger.

"That's a fucking krait," Joe said. He tried to pull it off with his left hand, but it wouldn't let go; it took him ten seconds to yank away the snake, a juvenile many-banded krait, and throw it back into the bag. He massaged the skin where he had been bitten and examined it carefully, but the puncture marks were too small for him to see.

Htun Win looked at him in consternation and told him that the snake had bitten him the night before, and he had experienced no symptoms. The others gathered in a knot to look at Joe's hand; they couldn't see where he had been bitten.

As they went in for breakfast, Dong Lin reached out and wrapped his big hand around the back of Joe's neck. "You gonna be all right, brother? I don't want you to die."

"Yeah. I'm gonna be all right." Lin stared at him intently. Joe shook his head. "I don't want to die either, bro."

They went inside and joined the other scientists at the camp's impromptu dining table, a cluster of child-size wooden school desks pulled together. As they breakfasted on Jiro's savory fried rice, the mood was subdued and serious but not grim. Joe joked, "My skin's thick. I don't think it got a fang into me. I'll be fine." He was encouraged by the fact that it was such a small specimen; if he had taken any venom, it wasn't a large amount.

After he had eaten, Joe told them he wanted to relax in his tent. As he walked away, he said, "I'll call you if I start to feel symptoms." It was seven o'clock.

Soon afterward, Jiro, unawares, came into Joe's room to ask him about the day's program. He found him lying in his tent and said, "Joe, are you OK? Why haven't you started working?"

Joe lay there, staring up into the top of his tent, picking at his right hand with his left. He was alone. He quietly told Jiro what had happened.

The guide was shocked. He asked fearfully, "Was it poisonous?"

"Yes, it was a krait."

Jiro had seen Joe wrangling venomous snakes all along the trail, and he had harbored a fear that Joe would be bitten. Wanting to be useful, he said, "Where is the knife? I will cut it and suck out the poison."

"No, it's too late. Jiro, can you try to get me a helicopter as soon as possible?" In Putao, they had discussed the possibility of a helicopter rescue, in case of an emergency. Then he said, "Can we find a respirator?"

"That would be very hard to do here," answered Jiro.

"Anyway, try for me."

Jiro ran out to the kitchen and sent three of his fastest men to the army base at Naung Mung, eight and a half miles away, to radio Yangon that the Americans in Rat Baw needed a helicopter immediately.

An hour after he was bitten, Joe felt a tingling in his muscles, and his hands began to twitch. He knew then that the bite had been envenomed. He called Lin over and asked him to tell the team to come into the room.

As his friends gathered around, Joe calmly explained what was happening to him. No one in the world knew more about the venom of *Bungarus multicinctus* than Joe Slowinski. He described the effects of a slowly deepening paralysis: The snake's venom works on several different parts of the nervous system simultaneously, blocking the nerve impulses that transmit instructions to the muscles, including those required to maintain life. There will be no pain, he told them. "First my eyelids will drop; I won't be able to hold them up." Soon he would lose the ability to speak and move his limbs, he said. Within a few hours, his respiratory system would shut down: The paralyzed central nervous system would be unable to instruct the diaphragm to breathe, causing a swift death by asphyxiation. He knew it would be impossible for him to reach the village hospital at Naung Mung, where they might be able to put him on a respirator before complete paralysis set in. The group decided to do what they could for him in Rat Baw. Perhaps a respirator could be brought from Naung Mung, but if not, they must give him mouth-to-mouth respiration until the helicopter arrived. Joe said, "You'll have to breathe for me."

If he survived forty-eight hours, he told them, the effects of the venom would abate as the neurotoxins worked their way out of his system; then his lungs might restart. Joe also warned them that if he was taken to a hospital, he must not be given antivenom injections. He told them about the severe allergic reaction he had experienced in South Dakota.

Everyone was remarkably restrained, at least in Joe's presence. They responded well to his cool, scientific analysis. Guin Wogan recalled, "Every person there said they would make this happen. We just told him, 'Whatever it takes.'"

Dong Lin asked Joe's permission to videotape and photograph events, in the hope of having a record of yet another amazing recovery from snakebite by Joe Slowinski. Joe agreed. Mark Moffett also took still photographs.

All work ceased, except in the kitchen. There, behind the schoolhouse, Jiro found Htun Win alone, looking dazed. Jiro asked him what happened: How had Joe gotten bitten? The young biologist, who idolized Joe, said that he had been confused. Yesterday, Pocho Soe Lwin had brought in two snakes that looked exactly the same. One of them had bitten Htun Win the night before, he told Jiro, which turned out to be the harmless Dinodon. Htun Win had assumed that both snakes were Dinodon, so he bagged them together. He was wrong. It was the worst mistake of his life: The other snake, the one that had just bitten Joe, was a krait. The young scientist began to wail, "Oh, no! Oh, no!"

The porters took the muddy mountain trail at a sprinting pace, and reached the army base at Naung Mung in less than three hours. By noon, the commanding officer at the base had conveyed their message by radio to his superiors in Putao; they relayed the news and request for help to Daw Marlar, who, as director of Thurein Travel, was Joe's official sponsor in Burma. She called the American Embassy and then the army; it is unclear whether she was calling the latter to request aid or merely to give them a report about the foreigner in her charge. At this point, Daw Marlar disappears from the record of the effort to save Joe Slowinski's life.

The consular officer on duty at the American Embassy, Elizabeth Jordan, took the call from Putao. She set in motion what she would later describe as the closest cooperation between the junta and the

Embassy in her three years in Burma. She immediately brought in Deputy Chief of Mission Karl Wycoff, the second-ranking American diplomat in Yangon. He had previously supervised a few medical crises, which had all concluded successfully. The last emergency had been in 2000, when a high-altitude balloonist who was trying to circumnavigate the globe crashed in the jungle. Now Wycoff began an extraordinary effort to get the Burmese government to waive its stringent rules against contact with foreigners and to cooperate in the attempt to rescue Joe Slowinski.

It was an unprecedented request: Wycoff was not only asking that the Burmese military send in one of its own helicopters to Rat Baw to pick up Joe and transport him to Myitkyina, he was also requesting permission for a medevac plane from Singapore, equipped with sophisticated emergency-care technology, to land at Myitkyina and take the patient. Wycoff had access to very high-level officers in the junta and kept hammering away at them to approve the mission.

"After nightfall," Wycoff said, "I went to their houses and knocked on their doors, but no one was there. Those guys didn't necessarily stay in their official residences." Unable to raise anyone at home, Wycoff went to the Ministry of Defense. "I camped out on their front gate. They ignored me for a while, so I got a tire tool from my car and banged on the gate until someone came down and talked to me." Then the negotiating began. The army had no problem with sending in the helicopter: They would charge $6,000 an hour for the mission, so it was a moneymaker for them. But no one had ever asked to land a private aircraft flying under foreign colors at Myitkyina.

Meanwhile, Jordan went ahead with preparations for the evacuation, so they could move quickly when the Burmese government gave the authorization. She knew it was quite possible that the army might approve evacuating Joe by helicopter from Rat Baw but refuse to allow the Singaporean medevac plane to land, so she called the head of the hospital in Myitkyina to ask him what medical supplies he would need if Joe were treated there. He replied, "We have nothing." They didn't even have a tracheotomy kit.

Jordan, unaware that Joe would refuse to take it, devoted much of the afternoon to a frustrating search for antivenom. She called every hospital, clinic, and medical supply house in Yangon, but there was none to be found. They had some antivenom at the American Embassy, but there was a strict policy against using their own medical supplies elsewhere, no matter how great the emergency might be. Jordan, a plainspoken Michigander, explained, "What if you give away the Embassy's antivenom, and then some kid at the Embassy gets bitten?" Yet when it became apparent that there was none for sale anywhere in the city, she decided to make an exception and brought the Embassy's antivenom.

As the morning wore on, Joe's physical condition deteriorated precisely as he had predicted it would. In stark contrast to the hysteria that prevailed after Joe was bitten by the cobra when he was filming with the National Geographic team, the scene at the schoolhouse in Rat Baw was wonderfully calm, even solemn. Joe lay down on his sleeping bag in his tent, with Moe Flannery and Guin Wogan lying next to him to provide human warmth and comfort. The men quietly gathered nearby. Joe asked someone to find an Ace bandage he could wrap around his right forearm to slow the traffic of blood and lymph in his hand, though by now the toxin had passed throughout his body. There was nothing more to be done except wait and see how serious the bite was.

When he began to feel numbness in his extremities, Joe kicked his legs and clenched and unclenched his fists, trying to stimulate circulation. "I'm having really weird sensations," he said. He drank a bit of *lokhu*. By ten o'clock he looked haggard and groggy as his eyelids began to droop. His voice grew hoarse and faint, but for as long as he could he continued to give his friends occasional reports on his body's reactions to the venom: "It's hard to breathe." "There's a pain in my stomach, a tingling sensation." "There are flashes of light when I lean back." It was as though he wanted to leave a scientific record of the event, if he didn't live to write about the experience himself. He and the team arranged a code of toe wiggles—one for yes, two for no—to use when he lost the power of speech.

It was a sunny morning but very humid, and the sandflies swarmed, as exasperating as ever. Outside, a cock crowed occasionally. Schoolchildren were singing and playing games in the soccer field opposite the school. Jiro sent half the porters into the village to wait.

Jiro and Joseph Tawng Wa, the schoolmaster, urged Joe to take *mashaw-tsi*, the local herbal cure for snakebite. The root is brewed to make a tea, and the leaf is chewed. Joe refused it, much to the chagrin of the Burmese, who believed implicitly in its efficacy. Joseph Tawng Wa claimed that no one in Rat Baw ever died of a snakebite, thanks to the miraculous curative power of *mashaw-tsi*. When Frank Kingdon-Ward passed through the region on his final trip to Burma, he identified the plant as a species of the genus Euonymus. At that time, a Kachin elder maintained a monopoly on the precious herb, gathering it in secret and selling it for exorbitant sums; but in 2001 it was widely available.

By ten forty, Joe's speech had become so slurred as to be incomprehensible. Everyone strained to understand what he was saying, but his voice was abandoning him; he needed all his power to breathe. He held up an eyelid with his finger, to maintain eye contact.

His friends tried to read his mind: "Do you need help, Joe?" "Need someone to breathe for you?"

Once he managed to murmur, "I have something in my mouth."

"Nice and easy, Joe," said Guin Wogan, and cleaned out his mouth for him.

Soon after that Joe uttered his last fully intelligible verbal expression: "I need . . . my mouth."

At that point Dave Catania began to give him mouth-to-mouth respiration. Both men wore the expedition T-shirt; the sweat from their beards commingled as the ichthyologist puffed air into his friend's mouth, filling his lungs. After a short period of artificial respiration, Joe resumed breathing for himself.

Although he could no longer speak, he was able to communicate by writing messages on a notepad, slowly, with a trembling hand. The first note was precautionary: "Not now, maybe later you'll have to intubate me, in my throat. Only when I say so. If I vomit it could be bad."

Around noon, Joe was no longer able to breathe lying down, so he sat with his back leaning against his friends, who were relieved by strong young porters. At one point, there were three people supporting him—American, Chinese, and Burmese. After a few minutes he scrawled, "Can I lean back a little bit." Then he wrote, "Please support my head, it's too hard for me." Dave Catania did as Joe asked and held a damp compress to his forehead. It was awkward trying to maintain a steady backrest for Joe with their bodies, so after a while they put a bench behind him.

At a quarter past one, Joe stopped breathing altogether. Moe Flannery, who was proficient in mouth-to-mouth respiration, started breathing for him. When she grew tired, she told the others between breaths that she would teach them how to do it. Every few seconds, she would cover Joe's lips with hers and blow air into his mouth; then she lifted her head and breathed herself, while Joe exhaled. Then she breathed for him again. Flannery, Catania, and Wogan assumed the principal responsibility for keeping Joe alive. Christiaan Klieger took responsibility for monitoring Joe's vital signs, measuring his pulse and heartbeat at periodic intervals.

In the beginning of the ordeal, the Harvard-Chinese contingent unobtrusively tried to be helpful, but as the vigil wore on, Joe's inner circle from Cal Academy took over. Dong Lin was always there. From time to time, he would put his arm around a member of the group and say, "I love you." The clicking of Mark Moffett's camera punctuated the long silences like the weird chirp of a mechanical insect.

Joe wrote, "No smoke." Jiro sent a porter to ask the Burmese to put out their cooking fires. Another porter fanned Joe with a square of cardboard.

At two twenty, the first of the couriers who had been sent to Naung Mung returned, reporting that the radio request for a helicopter had been relayed to Putao, where the military command, infuriatingly, asked for an update before they would send a rescue mission. The hopeful news was that a medical team was on its way from the hospital at Naung Mung. Jiro sent two fresh runners back with a more urgent message for Putao: The American visitor was now paralyzed and completely dependent on artificial respiration to stay alive. A helicopter was needed immediately. Jiro hopefully ordered the porters to spread a large square of red cloth in the soccer pitch in front of the schoolhouse as an impromptu helipad.

In the sickroom, Joe experienced a convulsive muscular spasm, which had the appearance of a reaction to pain. He tried to raise himself on his elbows, but couldn't. He wrote on the tablet, "I want to die."

Wogan chuckled once, resolutely, and said, "No! That we're not going to let you do."

Softly, Moe Flannery declared, "We're going to breathe for you, and you're going to get through this, Joe."

Mark Moffett, with understated intensity, said, "Kick some butt for us, Joe."

It was a moment of total, harmonious accord. The naturalists, in keeping with the spirit of Joe's rational calm, avoided hearty, bluff encouragement. Yet their powerful affection for Joe, their determination and intense desire to save his life, was a shining, manifest presence in the bare classroom.

Joe's last written message read "Let me di—": There, the pencil line trailed piteously down the page. He no longer had the strength to write.

It's a wrenching document, but it need not be interpreted as a cry of pain or spiritual anguish. Bryan Fry, the Melbourne-based herpetologist, said that one of the classic reactions to elapid venom is hypoxia: "There's a sense of disconnection, of sensory disruption—a floaty feeling." He said that when he was bitten by a death adder (*Acanthophis antarcticus*), a close evolutionary relative of the many-banded krait, "It was one of the most pleasant highs I've ever experienced."

Joe himself had predicted that he would experience no pain. It is even possible that his wish to be allowed to die was a product of the hypoxia produced by the venom. Perhaps the sense of disconnection and sensory disruption that Fry described allowed Joe to experience the scene in Rat Baw from a remote perspective, and to view death as an acceptable or even attractive option—a concept he would have abhorred in his right mind. Joe had an exceptionally powerful will to live, but now he lay near death; such radical ethical transformations, accompanied by a rapturous sense of joy, are commonly reported by people who have survived near-death experiences.

At five in the afternoon, spirits soared when a courier returned from Naung Mung with a respirator. A medical team was a couple of hours behind him, the runner said. Yet the scientists couldn't get the hand-operated machine to work: it was Asian-size, too small for robust Joe. His lungs were filling up with fluid, he was bleeding. It was getting messy.

The first member of the medical team to arrive was an old man. Bruce Bartholomew said that in China they would call him a "barefoot doctor"—in other words, he wasn't a doctor at all. He didn't seem to know much beyond how to measure blood pressure and pulse. He had no more success with the hand-operated respirator than the scientists, so the mouth-to-mouth respiration continued without a pause. The barefoot doctor wanted to give Joe an injection of penicillin, but Bartholomew, Klieger, and the others wouldn't allow it, because they knew it would do nothing to help him.

A short while later a real doctor arrived. Maung Maung Gyi was pale, skinny, and serious of demeanor; Jiro estimated his age at thirty-two, but he looked younger to the Americans. He appeared to know what he was doing, but he too was unable to get the respirator to work. It just wasn't big enough, and Joe's whiskers made it impossible to create an airtight seal, which was necessary to fill his lungs. All the machine accomplished was to scrape skin off Joe's face. Every minute that they fumbled with the contraption was a minute that Joe was deprived of oxygen. Inexorably, the toxicity in his bloodstream was climbing. Dr. Gyi, concerned about the buildup of fluid in Joe's trachea and possibly his lungs, suctioned out his throat, which helped stabilize his condition. There was little more that the earnest young doctor could do with his scant supplies.

Between six and seven in the evening, army officers from Naung Mung turned up with a portable two-way radio. They set it up in an unoccupied classroom on the northern end of the schoolhouse, at the opposite end of the building from where Joe lay. Radio communication was a circuitous process: The officers at the schoolhouse had direct contact only with Putao; information and requests had to be forwarded from Putao to Myitkyina, then relayed from there to Yangon. Anything the Americans in Rat Baw wished to say had to be translated into Burmese by Jiro. Error was built into the system, and in the early evening, heavy weather made the radio signal unreliable. Messages had to be repeated back and forth many times.

Twelve days later, Bruce Bartholomew wrote a detailed account of the events in Rat Baw in a report addressed to Patrick Kociolek, the executive director of the California Academy of Sciences. Here, Bartholomew set forth the team's grim assessment of the situation as the dark, wet night fell:

> During a break Moe came into the radio transmission room and said that she did not know how long she and the others could continue to keep Joe breathing with artificial respiration. She and I told Jiro to have the radio operator send out a message that unless the helicopter could arrive that evening and get Joe to a hospital in Myitkyina, we felt that Joe would not survive the night. I asked Dr. Maung Maung Gyi if he agreed, and he also said that he felt it was not likely Joe would survive the night.

The most ominous development was overhead: What had begun as the brightest sky of the expedition had turned into a lowering gray blanket. Until then, none of the scientists had doubted that a helicopter would arrive in time, but by nightfall it was raining steadily. Rat Baw was tucked into a rugged valley between steep ridges, which swirled with fleecy clouds even on a clear day: Safely landing a helicopter under the present conditions was out of the question. No one said a discouraging word, but it was shaping up to be a hellish night.

Martha Crow showed up at the office a bit early that morning, shortly before nine. A senior editor at *Food & Wine* magazine, she worked at the American Express Publishing offices in Midtown Manhattan, a few blocks south of Rockefeller Center on Sixth Avenue. As soon as she arrived, another editor told her that a plane had just crashed into the World Trade Center. Like many people, when Martha first heard the news she assumed it was a light aircraft that had strayed disastrously off course. She found a group of coworkers gathered in an office with a television and watched the appalling catastrophe taking place downtown.

Overwhelmed by the enormity of the event, as everyone in the electronic village was on September 11, 2001, Martha knew that she should return to her home in Brooklyn. But first, to help focus her mind, she took a deep breath and sat down at her desk to do a few odds and ends of work. The phone rang. It was Ron Slowinski, in Kansas City.

He said, "I have bad news," and told her about Joe.

Martha would later say that she thought she was losing her mind: Her external reality and her inner emotional life were disintegrating simultaneously. After Ron's phone call, she was on the verge of collapse. Some friends at American Express accompanied her home. They walked to the East Side and then headed downtown. The sooty smoke and filthy effluvia of the vanished skyscrapers hung over Lower Manhattan, a ghastly pall sickening body and soul. Martha walked past Union Square into Greenwich Village and eventually caught one of the last subway trains to Brooklyn that day. She retreated with her grief and terror into her brownstone just off Fort Greene Park, and wept.

When Ron Slowinski called his daughter, she had just sat down to do her Chinese homework in her Spanish-style bungalow in Silver Lake, Los Angeles. After Rachel made the first downward stroke with her brush, the telephone rang. Ron said, "Joe's been bitten by a krait." Rachel's initial thought was "This is it." She knew her brother would die. Joe had been bitten many times before, but this was different. By now, she knew, his expedition was deep into the sub-Himalayan wilderness, one of the most isolated places in the world. Yet she and her father told each other, "He's not dead, he's alive. He's been bitten before. Let's not jump to rash conclusions."

Ron's voice was somber, shaken. He said, "Of course, it's very difficult because of what's going on."

"What?"

"Yes, with everything that's going on."

"What do you mean? What's going on?"

He told her that two planes had flown into the World Trade Center in New York. When he added that the Pentagon had also been attacked, Rachel thought her father was suffering from some form of temporary insanity—that he was babbling, unhinged by the news of Joe's perilous situation. Then she turned on her television and saw a hole in the Manhattan skyline where the World Trade Center had been. When she lived in New York, in the early 1990s, Rachel had worked across the street from the twin towers: The place that used to be her office was gone now, reduced to rubble. Ron kept talking, but everything was becoming surreal. Rachel couldn't comprehend what was happening. She said good-bye and hung up the phone.

As soon as she collected herself, Rachel called her mother, anxious to hear her voice: Martha, she knew, belonged to a gym opposite the World Trade Center. It was immediately clear that Martha was terribly distraught. Rachel said what she could to calm her. She told her there was still hope, then hung up and had a good cry herself. Then she thought: I have to do something, Joe needs help. She started working the phone.

In Yangon, ten and a half hours ahead of Washington, the U.S. Embassy closed as usual at day's end. Liz Jordan moved the operation to rescue Joe Slowinski to her home in the landscaped compound maintained for American diplomats on Inya Lake, not far from the MiCasa

Hotel. Her husband, John Haynes, the Embassy's chief political officer, was helping her. As Jordan was preparing to call Ron Slowinski in Kansas City, where it was still early morning, Haynes shouted for her to come quickly to look at the television. They watched, transfixed with horror, as the World Trade Center tumbled down, the Pentagon burned, and ominous reports came in of other hijacked planes heading for Washington.

Soon after nine a.m. Central Daylight Time, just after the collapse of the north tower in New York, Jordan pulled herself away from the television to call Ron Slowinski. When Joe had registered at the American Embassy before leaving for Putao, he named his father as his emergency contact. Following the drill, Jordan called Ron to inform him that his son lay grievously ill in the jungle of northernmost Burma. That call set off the telephonic chain reaction in Joe's family, which eventually brought Rachel into contact with Jordan. The two women remained in close communication by phone throughout the dark night in Burma, the cataclysmic day in America.

Ordinarily, Jordan would have discharged her duty as consular officer once she made the initial call to the family in the United States; after that, the State Department in Washington would follow through. But conditions that day were as far from the ordinary as anyone had ever experienced. Washington was on a war footing. The American Embassy in Yangon was on its own, and they didn't blink. Karl Wycoff said, "Nine-eleven was a shock to us all, but we kept focused on the medical crisis before us. We were concentrating one hundred percent on Slowinski."

Under these exceptional circumstances, Liz Jordan was receptive to Rachel Slowinski's offer to help. Rachel had been to Burma, so she was realistic about what could be done. More important, she could provide the tactical support stateside that Washington, now under such extraordinary stress, could not. Jordan explained to her that the request to land the medevac plane from Singapore at Myitkyina had no precedent; it went completely against the grain of the paranoid, xenophobic military government. It wasn't like extracting a fallen hiker from Yellowstone: This was

Burma. However, they had one factor working for them: the junta's perpetual desire to curry favor with Washington. The leadership might be devoid of humanitarian impulses, but they would savor the prospect of having the Americans indebted to them for a favor—and saving the life of the famous scientist from California would be a big favor.

Finally, late in the evening, Karl Wycoff got approval for the rescue mission, all the way from the top, where every decision in Burma was made. "We were playing a little fast and loose with the Burmese," said Wycoff. "We still hadn't gotten the money squared away. But we figured with a major institution like the California Academy of Sciences, it would be all right." There would be plenty of time to sort the money out later.

A deal was struck. The Burmese Army would send in a helicopter to pick up Joe at Rat Baw, and the Singaporean medevac plane was cleared to land at Myitkyina. There, the two aircraft would rendezvous on the tarmac, and Joe would go aboard the flying hospital for emergency treatment and a return flight to Singapore, if necessary.

Liz Jordan got ready to fly out on the next scheduled flight to Myitkyina, which left just after dawn on the twelfth. She knew that in Myitkyina, even the most basic tools and technology might be lacking (or, more likely, broken), so she lugged a satellite phone with her, to stay in contact with the Slowinskis. It was the size of a laptop and very heavy. She didn't know how to use it, so she asked a neighbor in the American compound who worked for the Drug Enforcement Administration to come to her house and show her how. After she arrived in Myitkyina, however, she couldn't make it work, and had to use public telephones.

Rachel Slowinski described her frustration as "a scream that no one could hear. It was very hard to get a phone line in America. It was always busy." The American Embassy had accomplished a miracle by getting the generals to cooperate; now she had to deal with International SOS, the medevac company in Singapore. They demanded a $50,000 deposit before they would send the plane. The California

Academy of Sciences had insurance to cover an emergency such as this, but Rachel had difficulty making contact with them. There had been a two-alarm fire at the Academy on August 30, which had closed the museum to the public, and the City of San Francisco ordered all public buildings and offices closed on the morning of 9/11, as a precaution. The full horror of the attacks on the East Coast was known to most people in California before it was time to go to work, so few did.

As a result, Rachel couldn't get through to the Cal Academy administration. "There was no one at the museum, no one who could help Joe," she said. She called Jens Vindum at home, but he didn't know what to do either. Vindum tried to reassure her. He told her that Joe was probably better by now, feeding the denial that pulled hard at her, warring with an awful sense of doom. "I was panicking," Rachel said. "I knew it was very grave. I knew Joe was dying." All the while, she was triangulating her efforts with calls to her parents, seeking advice, trying to calm them and herself. She didn't think there was a problem about the money as far as the Academy was concerned, but events seemed to be conspiring against her. She was calling everyone she could think of who might be able to help her brother. "I was being a pill," she admitted.

By then, the army helicopter at Naung Mung was standing by, prepared to fly to Rat Baw, and Liz Jordan was on her way to Myitkyina.

Word of the perilous plight of the popular American scientist in Rat Baw soon traveled throughout the valley. By early evening, most of the people in the village had come to the schoolhouse compound. Joseph Tawng Wa said, "The women wept, and the men were very sad too." They milled around, pressing close with the bold, unabashed inquisitiveness of Asian country folk, to get a glimpse of the drama within, until Maung Maung Gyi told them to back off. The crowd drifted away to join the schoolteachers and the expedition's entourage. As the night wore on, more people arrived from farther out. By morning there were more than a hundred people gathered around the school, keeping a respectful vigil.

After nightfall, Joe's only means of communication was by the prearranged toe-wiggling code. Dong Lin said, "We asked him, 'If you hear us, wiggle your toes,' and he did, but he couldn't tell us what he was thinking." Joe was able to maintain contact in this way until the middle of the night. His last communication came after Moe Flannery, exhausted beyond all endurance, asked him if she and Guin Wogan could stop doing mouth-to-mouth and let Dong Lin take over. Joe's emphatic negative toe-wiggle indicated a strong preference for the women over his faithful friend. They interpreted it as a joke, a final jab of irony from the abyss of his paralysis. "He was such a macho man," Wogan said.

Later that night, a storm moved through the valley. The wind howled, pelting the tin roof with the strident staccato of rain, punctuated occasionally by startling peals of thunder. Joe's heart rate remained strong throughout the night and into the morning, but his disabled respiratory system was falling apart. When he exhaled, there was a loud rattle from his throat, and his fleshy lips spluttered, exuding a terrible smell of incipient decay. Some of the Burmese porters tried to give him mouth-to-mouth respiration, but they soon gave it up, unable to bear the smell. Joe's pale skin was mottled with livid bites from mosquitoes, sandflies, and leeches. At dawn on the twelfth, he was still alive, but he was steadily becoming less human.

By mid-morning his blood pressure began to fall. Dr. Gyi administered dopamine in a saline drip, to augment the power of the contractions of his heart. The drip raised his blood pressure for a while, but Joe's life force was sinking. At eleven o'clock, he went into shock. The doctor injected him with adrenaline to stimulate the heart, and the team continued to administer artificial respiration, but by noon Joe's purchase on life was a theoretical thread. His pupils were fixed and dilated.

Joe Slowinski died at 12:35 p.m., when his heart stopped. Twenty-nine and a half hours had passed since he was bitten by the many-banded krait. His heartbeat had been growing ever more faint; finally, it faded into nonexistence. Christiaan Klieger noticed the moment of death: The skin turned pallid, as quickly as a cloud passing over the sun dims a landscape. In the same moment, the leech and insect bites disappeared, as the fluids that had raised the bumps returned to the body's core. Klieger said nothing.

A few minutes later, the radio team brought the news that the helicopter had finally been able to leave Myitkyina. Moe Flannery began cardiopulmonary resuscitation, in the hope that Joe could be kept alive until the helicopter arrived; Dave Catania carried on with mouth-to-mouth respiration. At one o'clock, Dr. Gyi measured Joe's heart and pulse and quietly told Jiro in Burmese, "Everything is zero."

Jiro asked, "What does zero mean?"

The doctor replied, "It means death."

Bruce Bartholomew and Christiaan Klieger knew that Joe was dead, but the others refused to accept it. They continued to give him CPR, with Catania maintaining artificial respiration, for two more hours.

Bartholomew's memo to the executive director of Cal Academy tells the end of the scientists' heroic effort to save their leader from death:

> At slightly before 3:00 p.m. I went to find Dr. Maung Maung Gyi and asked him if in his medical opinion Joe was already dead, and he said yes, in his opinion Joe was dead, and had been dead since his heart stopped. At this time Jiro also told me that the radio operator had received notice that the helicopter had encountered bad weather and had to return to Myitkyina. I asked Dr. Maung Maung Gyi and Jiro to come with me to the room where Joe was, to make one more examination of Joe's vital signs. He checked for a pulse and shined a light into Joe's eyes to see if there was any reaction, but found neither. In front of all present I then repeated my question, did he consider Joe to be dead, and he replied yes, he considered Joe to be dead.

It was three o'clock in the afternoon. After a moment of stunned silence, the scientists streamed from the room, seeking places to be alone. The time for grieving had come.

Jiro reported Joe's death by radio to Putao, which relayed the news to Myitkyina. Liz Jordan, who had been standing by at the hospital there, called Ron Slowinski. He first phoned Martha, then Rachel. Martha didn't answer the telephone, so he left a message on her answering machine. "It was a ghastly thing to do," he later admitted, "but I just didn't know what else to do." Ron passed another sleepless, miserable night. In the afternoon, he looked out the window and realized that it was a gorgeous day, so he went to the neighborhood park where Joe used to catch turtles and snakes when he was a boy, and sat in the rose garden and cried for two hours.

Martha Crow, traumatized by the double catastrophe of 9/11 and Joe's snakebite, had managed to lapse into a comatose semblance of sleep by midnight. In the middle of the night, Rachel also tried calling her, but got no answer. Rachel was mad with worry because she hadn't heard from her mother, and above all wanted to prevent her from hearing about Joe's death from the answering machine; so she asked a friend of Martha's, whose daughter lived nearby, if the daughter could go check on her. Ron also phoned family friends who lived in New York, to ask them to get the police in Brooklyn to pay her a visit. Martha still hadn't checked her answering machine when a uniformed policeman knocked on her door at five a.m. to ask if her son had died in the World Trade Center. The question was garbled, like so many communications on that day, but Martha knew then that it was over.

Sandy Scoggin was in central Kansas at an international bird conference when Joe died. Late in the day on September 11, Sue Abbott, who shared the house in Marin County with Sandy and Moe Flannery, called her at her hotel when she heard the news. Sandy was devastated:

She was alone in a strange place, her country was under attack, and she had lost her love. With all civil aviation in America shut down indefinitely, she was stranded. When Sandy called Cal Academy for information about Joe, they gave her Ron Slowinski's telephone number; when she called him, he urged her to come to Kansas City so they could meet. She promised to come the next day.

She took the bus, crying all the way. After checking into the Holiday Inn, she called Ron. They met in front of Crown Center, Kansas City's bustling crossroads. When they found each other, they hugged for five minutes. Ron said, "It was so incredibly good to have someone to hug." Then they went to a café, where they ate and drank and talked about Joe. It was important for Sandy to hear more about Joe's life, to share the sense of family she felt for him.

After Liz Jordan received the news of Joe's death, she immediately shifted her attention to carrying out the painful, painstaking protocol mandated for the death of an American abroad, one of her principal duties as consular officer. Jordan prepared a long, detailed report of events for Washington. For all its dry, bureaucratic jargon, the document is infused with the sorrow and chagrin of those who had collaborated on the effort to save Joe:

> SUMMARY: AmCit Dr. Joseph Slowinski, leader of a National Science Foundation–funded scientific expedition into unexplored areas of northern Burma, died of snakebite on September 12, 2001. Post mobilized every resource to assist and evacuate Dr. Slowinski and prodded GOB [government of Burma] into providing an unprecedented level of support. Bad weather defeated our efforts, though, and Dr. Slowinski was cremated on September 13, 2001. . . . Post has met with the three AmCits who were evacuated with the body. The rest of the expedition had to walk out of jungle and is expected in Rangoon on September 21, 2001. End summary.

Name of deceased: Dr. Joseph Bruno Slowinski

DOB: 11–15–62

DATE OF DEATH: 12:35 p.m., September 12, 2001

PLACE OF DEATH: In the jungle near a small village called Rubbow [Rat Baw] in the Kachin States, northern Burma

CAUSE OF DEATH: Snakebite (Krait snake); no autopsy.

In Rat Baw, after the initial shock of disbelief had subsided, the Burmese men took over the care of the shattered wreckage of Joe's body. Htun Win and the rest of the field team washed it according to their tradition, and changed him out of his sweat-drenched shorts and T-shirt into long trousers and the best shirt they could find. Bruce Bartholomew asked David Boufford, the botanist from Harvard, to move the body to an empty classroom next to the impromptu radio room.

Bartholomew was assuming a position of authority, which annoyed some of the other naturalists. Quietly, they rebelled. Everyone was too exhausted and too fragile for a histrionic scene, but Joe's death didn't mollify the emotional strains that had existed among them before they arrived in Rat Baw.

Yet the living had to pull together now. They faced a grim array of options about what to do next. The first, hideous dilemma was what to do with Joe's body. Conditions couldn't have been worse: The prospect of carrying the body back to Putao was unimaginably gruesome, yet burial at Rat Baw without his family's consent seemed equally out of the question. They needed a helicopter now as much as ever, but the one that had been on its way to Rat Baw turned back as soon as the crew heard the news of Joe's death on the radio.

Liz Jordan, still at her post in Myitkyina, urged the regional commander to send it back immediately, but he refused. As she explained in her report to Washington:

> Although the military had permission to pick up Dr. Slowinski when he was still alive, they refused to pick up his dead body without per-

> mission from headquarters in Rangoon. The only alternative to a helicopter evacuation was for the team to carry the body out of the jungle—a five-day walk in 90-degree heat. ConOff [Consular Officer, Jordan herself] returned to Rangoon to try to obtain permission and discuss the disposition of the remains with the family. Again, the DCM [Karl Wycoff] worked tirelessly to persuade the GOB to provide a helicopter evacuation of the remains.

The radio report received by the scientists in Rat Baw after Joe's death that the helicopter had returned to Myitkyina because it had "encountered bad weather," which Bruce Bartholomew repeated in his memo to the Cal Academy administration, was later revealed to have been deceptive: In fact, the helicopter turned back because the crew did not want to risk bad luck by transporting a corpse on their aircraft. Jordan explained, "The Burmese are superstitious about putting a dead body in any kind of vehicle. I was really mad at them when they wouldn't send the helicopter back."

In the meantime, the scientists were fearful of immediate decay in the hot, damp atmosphere, which was teeming with insects. They had a large supply of formaldehyde with them for use on animal specimens; at their request, Dr. Gyi injected a liter of 10 percent solution into Joe's body, first in the body cavity and then the major muscle groups, thus partially embalming him. Bruce Bartholomew, Dave Catania, and Doug Long were present. The women, who had borne most of the burden of keeping Joe alive, asked to see him again, but Bartholomew said no. A military photographer took some pictures of the corpse. Then Jiro supervised wrapping it in a plastic sheet, leaving only the face open, with a mosquito net to cover it.

In the evening, the group had its first hot meal in two days. They discussed the other major decision confronting them now: What to do with themselves? Dong Lin, crazily, said that the expedition should go ahead, that that's what Joe would have wanted; but Klieger, who was Lin's best friend among the surviving group, immediately vetoed the idea. He said, "This expedition is over." No one disagreed with him. All collecting ceased after Joe's death.

That night, Dave Catania revealed to the others what he had known for many hours. In the middle of the night before, when he went into his tent to rest for a moment from the exertion of giving Joe mouth-to-mouth respiration, he listened to his shortwave radio and heard the news about the attacks on New York and Washington. Catania had kept it to himself, waiting for the right moment to tell his colleagues the shocking news. When he finally told them, in his subdued way, what had happened, they couldn't grasp it. Klieger said, "It was like he had said Jupiter collided with Saturn. It was just incomprehensible."

The scientists were too overwhelmed by the palpable reality of Rat Baw to accommodate any theoretical reality in the outside world, which now seemed farther away than ever. When they had a chance to take in the news, and as they listened to more broadcasts, like many people at that time they wondered if it might not be the beginning of Armageddon. Some of them asked whether they might be wise to stay put: They had plenty of supplies, and Upper Burma would be the last place to be targeted in any counterattack.

In Yangon, Karl Wycoff went back to the general who had approved the earlier mission and got him to order the helicopter to return for Joe's body. However, the weather on the twelfth was too changeable all day for the helicopter to fly to Rat Baw. Situated as the village was, in the lap of a deep valley, it was impossible to see more than half a mile in either direction, so predicting atmospheric conditions even a quarter hour in advance was little more than guesswork. The command in Myitkyina was dickering: It would be much easier for the helicopter if the Americans could transfer the corpse to the army base at Naung Mung. That seemed reasonable to the botanists, but the others resisted: Joe's body would not be moved.

The mission was scrubbed until the next day.

On September 13, the weather system throughout the region cleared, and a firm pickup time was scheduled for the early afternoon. The embalmed body was already sealed in plastic and trussed up with duct tape, but, in a macabre twist, an army officer said they needed additional photographs because Joe's face had been covered by a piece of cloth in the pictures taken before. Bruce Bartholomew unwrapped Joe's head for the photographer to reshoot it.

There would be room on the helicopter for four expedition members to leave with Joe's remains; the rest would have to walk out. While Joe was still alive, and they were hoping to evacuate him, the group had decided that Wogan, Flannery, and Htun Win would accompany him—the women to maintain artificial respiration, and Htun Win as the senior Burmese representative. Mark Moffett was going with them; he had a professional obligation in Malaysia, he said. He had been planning to return to Putao on the morning of September 11, but obviously decided to stay on after Joe was bitten.

Now Bartholomew suggested that one of the American men should accompany the body, and it was agreed that Moe Flannery would remain behind to make room for him. The group favored sending Christiaan Klieger. The anthropologist was unencumbered with specimens and, more important, as Cal Academy's grants officer, he was the expedition member closest to the museum's administration. Klieger refused. He said, "I have to walk out." He wanted to be with Dong Lin: The volatile photographer had been closer to Joe than any of the others, and Klieger thought he needed someone with him. So the group decided to send Doug Long.

The choices seemed right. Wogan was levelheaded and competent to deal with the trial of transporting Joe's remains. It was essential to get Htun Win out, to prevent the group from splintering, if not exploding, over recriminations about his role in the snakebite; anyway, his sense of shame made him difficult to be around, even for those who were sympathetic. Doug Long was the slowest walker, and no one would miss Mark Moffett.

The porters prepared a landing place for the helicopter. After discussing it with the soldiers, they cut some trees and laid them out in a circle large enough for the helicopter to land in, in the soccer field in front of the school. In the center of the circle they made a small cross, with wood ash and red clothing borrowed from the porters, to serve as the target. At the far corners of the field, soldiers set smoky fires to help guide the pilot; armed soldiers, following regulations, occupied the perimeter.

When the helicopter came into view at the southern end of the valley, porters swiftly carried Joe's body out of the classroom where it had lain since its removal from the death room. After the helicopter landed, they gently loaded him into the bay. Then Guin Wogan, Doug Long, Mark Moffett, and Htun Win scrambled aboard. At three p.m., September 13, Joe Slowinski finally left Rat Baw.

As soon as the helicopter landed in Myitkyina, Doug Long brought Joe's body to the city morgue and had it packed with ice, in the logical expectation that he would accompany it to Yangon, where it would be transported on to America, to Joe's family, for burial. Long called Liz Jordan, who was still in regular telephone contact with the Slowinski family. He was enormously relieved when she told him that they had decided to have Joe's remains cremated in Myitkyina. Jordan was frantically busy dealing with the American citizens in Yangon whose lives had been upended by 9/11, so her husband, John Haynes, was on his way to Myitkyina to assist in the formalities. However, Haynes would not arrive there until late, after the last flight to Yangon had left. Mark Moffett and Doug Long decided not to remain in Myitkyina for the cremation, and took the next flight to Yangon.

Upon their arrival, Liz Jordan debriefed them at their hotel. In the course of her meeting with Moffett, he made a surprising statement. In her report to Washington, Jordan wrote:

> [Moffett] told ConOff that although he had applied for a Burmese visa claiming to be a biologist, he was actually a journalist for *National Geographic* magazine. As soon as the snake bit Dr. Slowinski, he began documenting all of the events . . . He said he would submit the story to *National Geographic* magazine as soon as he returned to the States. He was leaving Burma that evening and was concerned about being able to get his film out of the country. ConOff advised him to call if he encountered any problems at the airport, but we never heard from him.

Moffett left Burma as soon as he could get a flight out. (*National Geographic* didn't publish Moffett's story about Joe's ordeal; it appeared in *Outside* magazine, in April 2002.)

Doug Long, now on his own in Yangon, wrote a long, passionate e-mail describing Joe's last days, addressed to the staff of the California Academy of Sciences. It was filled with inaccuracies, including a highly implausible description of the fatal snakebite, which would later cause some confusion. Nonetheless, the letter filled the urgent need in San Francisco for some basic information about Joe's death. In conclusion, Long explained, "We decided that Guin Wogan, Joe's graduate student, department secretary, and veteran of previous Myanmar expeditions, and Htun Win, Joe's beloved right-hand man in the field, would stay another day in Myitkyina to oversee the cremation and accompany the ashes back to Yangon."

In Burma, human remains are usually cremated on a wooden pyre, but John Haynes told the authorities that that was out of the question. The ceremony took place at a gas-powered crematorium, old but clean and well-maintained, in the countryside outside Myitkyina, next to a graveyard that was open to all faiths. Misty blue mountains shimmered faintly in the distance.

It was a sunny, cool day. Present were Haynes, Guin Wogan, Htun Win, and the regional army commander, whom Haynes described as "a nice, affable guy." Haynes talked to Wogan while they waited for Joe's ashes. She was in bad shape, he thought, obviously exhausted and traumatized. Her face was chapped and rubbed raw from the hours of giving mouth-to-mouth respiration to a bearded man. Haynes said, "Part of what I was doing there was to make sure that Guin was OK, to comfort her, to keep her mind off the bad memories." She told him how the snakebite had occurred, and about the rest of the ordeal at Rat Baw. "Joe was always a risk-taker," she told him.

It took nearly three hours for the fire to consume the corrupt remains of Joe Slowinski: It was a clean job, although some big chunks of bone remained. At the end, the ashes were put in a jar, which the officer wrapped in a white cloth. Then the Burmese men—Htun Win, the army officer, and the crematorium's employees—performed a Buddhist ritual.

Haynes brought the ashes back to Yangon and gave them to Liz Jordan, who turned them over to Christiaan Klieger when he returned. She conveyed Ron Slowinski's wish that his son's remains be kept in a lacquer bowl of the sort used by Burmese monks for begging food. Klieger and Moe Flannery bought such a bowl, poured the ashes and bone into it, and sealed it. Before his flight home, Klieger gave the bowl directly to the director of Thai Airways in Yangon.

When the box containing Joe Slowinski's earthly remains came bumping down the baggage chute at San Francisco Airport, all the senior staff members of the California Academy of Sciences were there to receive them. Patrick Kociolek collected Joe's ashes and kept them in the Academy safe until Rachel Slowinski took possession of them.

Six years later, she and her parents still hadn't decided what to do with them.

Jiro needed a full day to organize the logistics for the return trek to Putao, so the remnants of the expedition spent two more nights in Rat Baw after the helicopter left with Joe's body. Jiro had already bought the meat for the barbecue Joe had planned for September 11, and the group decided to go ahead with it. It would be a public wake: The porters and the workers in the village who had assisted the expedition would all be invited. Joe had planned the party in detail; even after he had been bitten, before the effects of the venom set in, he discussed the menu with Jiro. There would be a great feast in Joe's honor on the expedition's last night in Rat Baw.

The cooks worked all day, grilling the meat on bamboo skewers, and porters were sent to Naung Mung to buy rice wine. The party got off to a slow start. When the naturalists invited the porters and other workers to join them in the little campus in front of the schoolhouse, they were too shy to join them. Everyone felt a bit awkward. Then a fire was built, and benches were brought out from the classrooms. A hundred people sat around the fire, eating beef barbecue—a supreme luxury in impoverished Rat Baw. The hosts passed out bags of candy from the expedition's stores. After the wine was poured round, the porters sang traditional songs in the Rawang language. They sang a song of prayer for Joe.

The drinking and singing went long into the night.

Epilogue

Throughout his life, Joe Slowinski craved official recognition; in death he received accolades of the sort accorded to the greatest names in his field. The Center for North American Herpetology in Lawrence, Kansas, established an essay competition in his name, an annual cash prize for the best published paper in the field of snake systematics. The award was supported in part by a generous donation from Sean Windsor, Joe's best friend in high school. The California Academy of Sciences created a Joseph B. Slowinski Memorial Fund, which provides cash grants for field research in herpetology.

Frank Burbrink named a new species of snake after his mentor. *Elaphe slowinskii* (later renamed *Pantherophis slowinskii*), the first new snake species to be identified in North America since 1949, is a corn snake, so called because its black-and-white speckled belly resembles maize. A handsome nonvenomous colubrid, Slowinski's corn snake reclusively inhabits parts of Louisiana, Texas, and Arkansas. Burbrink actually chose the name a few weeks before Joe's death. "I decided to name *Elaphe slowinskii* in August 2001, out of pure appreciation," said Burbrink.

"I caught my first corn snake with Joe. We found it in the wild, on the road, somewhere in Mississippi."

Herpetologist Aaron Bauer named a bent-toed gecko from Burma after Joe: *Cyrtodactylus slowinskii*. And in 2005, Ulrich Kuch, at Johann Wolfgang Goethe University in Frankfurt, named a new species of krait discovered in the Red River valley of Vietnam, near Hanoi, for the scientist who had done so much to rationalize the phylogeny of Asian elapids: *Bungarus slowinskii*.

In Burma, Joe's friends mourned him with prayers and offerings at their shrines, then quickly moved on to the business of living—a typical Asian response to death.

The California Academy of Sciences attempted to follow the Burmese example. The museum decided to mount a handsome memorial service in Joe's honor, and then declare an end to mourning. What was the use of picking away at the scar of a wound that would take a lifetime to heal?

Thus the Academy never carried out any organized investigation to determine what had gone wrong at Rat Baw. Not that it would have been easy to do: in the emotionally turbulent aftermath of Joe's death, the expedition members' recollections of the exact sequence of events leading up to the fatal snakebite became confused and contradictory, to a degree that mystified and confounded everyone involved.

In his article for *Outside*, Mark Moffett reported that Htun Win said, "I think it's a Dinodon," then Joe put his hand in the specimen bag. However, in Dong Lin's recollection, Joe hesitated and asked Htun Win a second time if he was sure the snake in the bag was nonvenomous, and Htun Win replied, "Sure, sure, sure." It is more plausible that Moffett either forgot Joe's second question or omitted it to streamline his narrative than that Dong Lin invented it; yet Lin himself has affirmed that Moffett's account of the episode is essentially correct, a judgment supported by other members of the expedition.

It might seem to make a big difference: Surely if Joe asked a second time, and was again reassured that the snake in the bag was nonvenomous, then his action was less foolish, and Htun Win's culpability the greater. Yet herpetologists who knew Joe well vehemently disagree. It is almost unimaginable, they say, that any skilled scientist handling a collection of snakes that included so many venomous species would put his hand into a specimen bag without first eyeballing its contents. Unimaginable, perhaps, but that's what Joe did.

There were still other versions. The obituary in the *San Francisco Chronicle* reported that Joe *did* look into the bag before he put his hand in it, a scenario at odds with all eyewitness accounts. In the *Los Angeles Times*, the obituary quoted Amy Cramer, a spokeswoman for Cal Academy, who stated that "based on reports e-mailed from researchers at the remote site," the snake that bit Joe was the same snake that had bitten Htun Win. That is theoretically possible: Htun Win could have received a dry bite. Yet it's far more likely that he confused two snakes—the Dinodon that bit him and the many-banded krait that killed Joe.

Both of those unlikely narratives were based on the e-mail message that Doug Long sent Patrick Kociolek as soon as he arrived in Yangon, which was riddled with errors. Yet a fifth version, quite different from the others, came from an eyewitness. In an interview with Roy Wenzl, a reporter for the *Wichita Eagle*, Guin Wogan said that when Joe asked Htun Win what was in the bag, he answered, "I don't know." This seems even less plausible. It is almost unimaginable that Joe, however hungover and absentminded he might have been, could have put his hand in the specimen bag with absolutely no information about its contents. Yet no witness could be more trustworthy than Wogan: She was one of Joe's best friends and a close professional collaborator. (Subsequently, she has said that Roy Wenzl misquoted her, but she has declined to discuss the matter further.)

These conflicting accounts angered Dong Lin, Joe's closest friend among the expedition members. He was particularly incensed by the article in the *Wichita Eagle*, which was one of the most substantial newspaper stories about Joe's death. He believed that the museum administration set a higher priority on preserving the collaboration with the Burmese than on protecting Joe's memory, and that it was orchestrating a campaign to absolve Htun Win of blame. "The Academy just wants to forget

Joe," Lin said often. He furiously tried to recruit others to his camp, but the scientists, mostly a temperate lot, were ill-equipped emotionally to deal with the growing atmosphere of passionate divisiveness.

Things came to a head at Joe's memorial service, which was held at the Academy on Friday, November 9, 2001. Hundreds of people turned out. The ceremony itself went well. Administration officials spoke warmly of Joe and the heartbreaking sadness of his loss; Robert Drewes evoked fond memories of their standing handball date. Martha Crow spoke about Joe's childhood; her twin, Mary, read a poem. Stan Rasmussen and Frank Burbrink flew in to reminisce about their friend's youthful exploits. Jay Savage, now in retirement, lightened the somber mood by telling the story about Joe in the nudist resort. Rachel prepared a slide show of photographs of her brother from every period of his life; Dong Lin edited a video of Joe in the field in Burma. Poster-size enlargements of Lin's portraits of Joe hung on the walls.

At the conclusion of the program, cocktails and sushi, one of Joe's favorite foods, were served. It was a lavish, tasteful event, but the atmosphere was awkward: too solemn and restrained for a wake, yet too sentimental to be anything else. As the event wound down, Joe's friends from the Academy migrated to the herpetology lab in the basement, where they met every Friday at day's end for beer and banter. Joe had always presided over these gatherings, unless he was in Burma.

The impromptu get-together was a disaster. Dong Lin became more vociferous than ever about his belief that Htun Win should be blamed for Joe's death. Then he turned on Guin Wogan, whose comments to the *Wichita Eagle* had seemed to exonerate the Burmese scientist. They got into an argument, which quickly overheated. Jens Vindum intervened, taking Wogan's part. It even appeared that the quarrel might turn violent; at that point, the group ordered Dong Lin to leave. After this debacle, he was virtually ostracized.

In Burma, too, Joe's legacy appeared to be disintegrating: The Myanmar Herpetological Survey teetered on the verge of dissolution. Joe had been the project's vital force, its intellectual fountainhead and inspirational motivator to such an extent that no one could imagine it going on without him. The conjunction of Joe's decision to go into the field without the sponsorship of the Forest Ministry, the circumstances of his death, and Dong Lin's accusations against Htun Win made the Burmese partners in the project skittish, and the Americans fearful about its continuing viability. George Zug said, "The whole project looked like it was unraveling. Every time we sent an e-mail to Burma, it was misinterpreted."

When U Khin Maung Zaw informed his American collaborators that he intended to replace Htun Win, in response to Dong Lin's attacks on him, they took action. Zug and Jens Vindum, who was doing his best to fill the vacuum at the Academy, strongly believed that Htun Win was the project's most valuable asset now and could provide the leadership it desperately needed. In December 2001, the two Americans flew to Yangon, essentially to save the project. They met with U Khin Maung Zaw and made a strong pitch for putting Htun Win in charge. Zug and Vindum assured Zaw that in any future expeditions, rigorous and orthodox standards of safety would be enforced. They promised that they would never again go to an area that the Forest Ministry declared off-limits—no arguments, no resorting to other ministries. As a conciliatory gift, Vindum brought Zaw a new camera. The effort succeeded; the collaboration stabilized.

Most of the responsibility for field exploration now fell on the Burmese, under Htun Win's supervision. Scientists from Cal Academy concentrated on staff training and the establishment of the new natural-history museum in Hlawga Wildlife Park. There were occasional forays: In 2003, a Cal Academy group with Vindum, Wogan, and research associate Jeffery Wilkinson ventured to Kachin State to conduct a survey of the herpetofauna in a new wildlife sanctuary at Indawgyi Lake.

In 2004, for the second time, a project leader was tragically struck down when Htun Win died of AIDS. He was thirty-one. Troubles now assailed the Myanmar Herpetological Survey from every direction. The

U.S. government imposed more stringent trade sanctions against Burma, which for a time prevented the American scientists from bringing their specimens back into the country, on the grounds that they were imports; then funds from the National Science Foundation dried up. One by one these hurdles were surmounted, and by 2007 the project was moving forward again, with a successful field expedition in August to the northern Pegu Mountains.

Yet soon after team leader Jens Vindum left Burma, monks in Yangon took to the streets in a series of mass pro-democracy protests, which were suppressed by the army in the nation's most brutal crackdown since the student massacres of 1988. Once again, the Myanmar Herpetological Survey was thrown into chaos.

Anyone who ponders Joe Slowinski's life and death gets emotionally involved and inevitably asks the question that haunted Dong Lin: Who, finally, was responsible for what happened in Rat Baw?

In the first analysis, if anything is certain it's that no one who played a part in the attempt to keep Joe alive and rescue him can be faulted. It was a heroic effort, which brought out the finest in everyone involved.

Some of Joe's friends have suggested that if Alan Rabinowitz hadn't given him a bad review at National Geographic, then Joe might have gotten a grant from the society and had enough money to mount a better, safer expedition. Yet if Joe had had unlimited funds, it's doubtful that he would have made very different arrangements. Little more could have been done to improve the conditions on the trip: The miseries were elemental, the failures human. It wasn't lack of money that kept the helicopter from getting to Rat Baw in time, it was the weather. A two-way radio would have done more than anything else to facilitate a rescue operation, but the Burmese government forbade the Americans from bringing one with them.

Daw Marlar must be seriously faulted for failing to honor her commitments. Joe's journal is replete with grievances about her duplicity and broken promises; other expedition members afterward complained bitterly about her slipshod performance. In an interview in Yangon, in 2005, Marlar even claimed to have had nothing to do with the expedition. "I had no experience with the travel business," she said, and added, absurdly, that she barely knew Joe Slowinski.

Yet it was the team leader who decided to go ahead after it became clear that Daw Marlar wasn't going to fulfill her promise to provide medical personnel. As damning as that might seem in hindsight, it is not a usual practice to bring doctors on expeditions of this sort: A certain level of risk must be assumed by the participants. It is inherently dangerous work. Even if the promised medical professionals had materialized, it's an open question whether they could have saved Joe without a functional respirator to sustain his life until the potent toxin worked its way out of his system. It was a prodigy of human endurance on Joe's part, and of devotion on the part of his friends, that his "two steps" lasted nearly thirty hours. If the weather had remained fine and the helicopter had been able to reach Rat Baw on the eleventh, if the medevac from Singapore had made it to Myitkyina in time, then Joe might have been saved—or he might not.

Ultimately, the person responsible for Joe Slowinski's death was Joe Slowinski. He was the boss. There was no one in San Francisco looking over his shoulder when he made the preparations for his expeditions, and to the Burmese he was the lordly bearer of gifts, a distinguished foreign guest exempt from open criticism. Joe had built up a formidable power base: No one presumes to tell a scientist who has raised millions of dollars for his research what to do with the money. Since the time of Humboldt, the leader of a scientific expedition has been regarded as something of an intellectual prince, granted almost unlimited autonomy.

There was no doubt that Joe took full advantage of the wide latitude granted him. His sunny natural charm was his secret weapon: When he was in good form, bounding with enthusiasm, no one could say no to him. Even when he pushed the outer limits of discretion, he was always guided by a solid sense of fair play. The ethics of certain aspects of Joe's strategy for the Hkakabo Razi expedition were questionable, no doubt, but they all had precedent and fell within broadly accepted parameters—or at least in a gray area. In his personal conduct, there were plenty of "warning signs," if you chose to look at it that way: Tales of his reckless snake-handling were legendary. Yet that, too, was within the expected range of behavior for a field herpetologist. A love of danger was acceptable, even tacitly admired—as long as your luck held.

The 2001 expedition, however, presented daunting challenges at every turn. Joe had had plenty of experience with leeches and malaria and heavy rain, but the trail conditions in northern Burma in September 2001 took misery to a new level, which frayed his nerves and vitiated his mental powers. Joe was accustomed to working with friends, his inner circle from the Cal Academy herpetology department and his trusted Burmese field team. Instead, he was in charge of fifteen scientists with competing interests, many of whom he barely knew. Afterward, several team members agreed that the pressures of trying to maintain harmony among the expedition's fractious participants under extremely arduous circumstances took a heavy toll on Joe, one that was exacerbated by too much alcohol and too little sleep.

Yet there would have been no expedition to Hkakabo Razi if Joe hadn't dreamed it, and no one else could have led it. It was more than that: Joe was *compelled* to lead the expedition. Nothing could have stopped him. The opposition of another prominent scientist, the rejection of a major grant application, the denial of sponsorship by his usual Burmese collaborators—these obstacles only strengthened his resolve until, one by one, they fell away in the face of his seemingly indomitable will. The tragedy of Joe Slowinski might be that his determination in the face of adversity contributed to conditions so stressful that they eventually weakened his judgment; and when the big expedition he had been planning for years started to fall apart, he lost his ability to cope. All his life, Joe Slowinski had struggled to shape events to his vision, but when he reached into that specimen bag, he surrendered himself to chance.

Even if he never talked about it, Joe always knew that the way he lived and worked might someday be the death of him, but he wanted no other life. He believed that the world we live in is a product of random events; his death was one of them. For all the creative intelligence encoded in Joe Slowinski's genes, they would not be passed on. His afterlife would be in the literature of biology. *Bungarus slowinskii, Cyrtodactylus slowinskii, Pantherophis slowinskii*: these are his progeny.

The ongoing census of terrestrial life forms, given its skeleton by Linnaeus and its philosophical flesh by Darwin, is one of the boldest intellectual collaborations in history. For centuries, scholars and amateurs throughout the world have labored mightily and endured terrible hardships to compile a comprehensive portrait of life on Earth, all of them playing by the same rule book—a vast, open document that is in a constant state of revision. It's the biological equivalent of a Grand Unified Theory: a verifiable, logically sound explanation of everything. As unimaginably complex and difficult as the task is, for the insatiably curious among us it possesses an irresistible allure—precisely because it can be done.

The vast majority of humankind will live their lives without giving one thought to the Burmese spitting cobra. Joe's lucid, masterly description of the affinities and divergences among the elapids of Asia will never yield a material payoff: no useful new technology, no advance in agricultural productivity, no miracle cure. Yet Joe Slowinski's work will live as long as science does: It forms an irreplaceable passage in the great epic poem of life being composed by biologists, one of the most profound and elegant accomplishments of our species.

Sources and Methods

My narrative of the life and death of Joe Slowinski is completely factual, in the sense that every sentence is based upon facts: documents in the form of journals and letters, my interviews with eyewitnesses, previously published accounts of events, audio and video recordings, and my own observations of people and places. Where there were gaps in the record, I was silent. Every subjective description of people and events is based upon a direct observation by a participant; wherever possible, I have told the story using the words of people who were there.

I frequently found contradictory versions of the same event; the widely conflicting accounts of the fatal snakebite are only the most dramatic example of the phenomenon. In those cases, I followed the accounts of those who seemed to me the most reliable and impartial witnesses. I don't believe that any of the approximately seventy friends and colleagues of Joe Slowinski I interviewed while researching this book ever intended to deceive me, but it is human nature to see events as confirming one's own beliefs and hopes, and the passage of time smoothes away disparities and richly colors events that have faded into insubstantiality. Many key events are reported by only one participant—an inevitable consequence of writing a book about a deceased person. Undoubtedly there are some inaccuracies here as a result, but everywhere I have used the best information at my disposal.

My narrative of the expedition to Hkakabo Razi is based primarily upon Joe's own, very detailed journal, written at the end of each day, and intensive interviews with several members of the expedition. Other points of view come from journalistic accounts reported and written soon afterward: newspaper stories by Edward Eveld, in the *Kansas City Star*, and Roy Wenzl, in the *Wichita Eagle*; Robert Sullivan's brief feature in the *New York Times Magazine*; and Mark Moffett's memoir of the trip in *Outside*. The descriptions of the landscape and people of Upper Burma are reflections of my visit to the region in January 2005. My description of Joe Slowinski's death is based upon Elizabeth Jordan's official report, written immediately afterward; Bruce Bartholomew's finely detailed account, dated September 23, 2001; and other records of the event, in various media.

I was able to make contact with almost everyone who played a major role in Joe's life at every stage, and nearly all of them gave generously of their time and resources. A few of Joe's colleagues at the Academy chose not to talk to me, but most of those who were closest to him there have been cooperative and helpful. A fuller account of my sources may be found in the acknowledgments.

My most invaluable resource in researching Joe Slowinski's life was Joe himself. He was a conscientious journal-keeper, a confirmed printer of e-mails, and a pack rat—all endearing qualities to a biographer. He was rarely reflective in his writings, which are now archived at the Cal Academy library, but he was always vivid and clear, and lovable even when he was exasperating. As I was working on the book, I keenly regretted the occasional gaps in his journal. Now that I've concluded the telling of his story, it's the great gap at the end I regret most of all. It's a peculiar kind of sadness to feel sorely the loss of someone I never met.

A Note on Place Names: Burma versus Myanmar

Some American readers may be surprised to find that the country where many events in this book take place is called Burma, for most American news organizations (though not all) follow the country's military government and use the name Myanmar. The latter is a recent coinage, of specious origin. In 1988, the State Law and Order Restoration Council (SLORC), the military junta governing the country, established a Commission of Enquiry into the True Naming of Myanmar Names. The group had twenty-one members: four scholars (two specialists in the Burmese language and two in English), eight army officers, and nine civil servants. Charged with creating English transliterations that more closely matched the Burmese names, the commission proposed sweeping changes: In addition to calling the country Myanmar, the capital city was renamed Yangon, from Rangoon. Other alterations were purely orthographical—the nation's great river changed from Irrawaddy to Ayeyarwady, the famed archaeological district of Pagan became Bagan, and so forth.

The policy attracted little attention inside the country, but outside its borders the edict was controversial from the start. "Myanmar" has few adherents outside the Burmese government, the United Nations, and the American media. In Great Britain and the Commonwealth nations, with the exception of Canada, "Burma" is still the norm. (No changes have been proposed for any other languages; the French still speak of Birmanie and the Spanish retain Birmania.)

The change was obviously political in motive: The Commission of Enquiry into the True Naming of Myanmar Names justified its decisions by avowing the purpose of ridding the country of false, British-imposed pidgin names. However, most of the English-speaking Burmese people I have met call their country Burma and avoid "Myanmar" as a creature of SLORC (which changed its name to the State Peace and Development Council in 1997). The Burmese may have resented and even hated their imperial British overlords, but they loathe their present government with a ferocious intensity.

The National League for Democracy, the main opposition to the military junta, calls the country Burma in its publications; the major government-in-exile groups are the National Coalition Government of the Union of Burma and the Members of Parliament Union of Burma. The U.S. Department of State calls the country Burma as a matter of policy.

While there are sound arguments for following the name imposed on a place by its government, whether or not one likes the people running it, the only persuasive reason I can see to choose "Myanmar" is to be in harmony with the copy desks of influential American news organizations. However, in this book I have followed the country's democratically elected government-in-exile and my own country's government, and use "Burma" and the other pre-SLORC place names.

The exception to my rule is that I have called the nation's capital Yangon, following the usage universal among the Burmese people I have met; "Rangoon" truly seems to be a colonial relic.

I hope that these choices will not offend any civilian citizens of that beautiful, unhappy land. Everyone else is, of course, welcome to disagree with me.

Acknowledgments

This book could not have been written without the cooperation and support of Joe Slowinski's parents, Martha Crow and Ron Slowinski, and his sister, Rachel Slowinski. They have been exceptionally generous and helpful. They gave me full and unrestricted access to Joe's journals and private papers, even mementos of his childhood; more important, they admitted me to the private places in their memories of their son and brother. If this book comes close to meriting the trust they have placed in me, I shall have accomplished one of my principal aims.

Joe Slowinski was an extraordinarily well-loved man. I've been very fortunate in the warm, open-hearted response I have received from nearly all his friends and colleagues whom I approached for help. Above all, I am profoundly indebted to Brian Crother, Christiaan Klieger, Chris Wemmer, Wolfgang Wüster, and George Zug, eminent scholars all, who were unfailingly gracious and forthcoming when I returned to them again and again with more questions about Joe's life, more requests for scientific information. They read extended excerpts of the manuscript and made many important factual corrections and excellent suggestions for improving the book's overall accuracy. Other scientists who made substantial contributions of personal reminiscence and insight into Joe Slowinski's intellectual development are Brady Barr, Frank Burbrink, Henry Fitch, Craig Guyer, Larry Martin, Kevin de Queiroz, Roy McDiarmid, Jay Savage, Carol Spencer, and Jens Vindum. Joe Collins, Maureen Donnelly, and Travis Taggart were also generous and thorough in their responses to my queries.

I'm deeply indebted to the members of Joe Slowinski's final expedition who shared intimate memories of that terrible event: Christiaan Klieger, Dong Lin, and U Thein Aung were especially generous, setting forth their recollections in careful detail. Bruce Bartholomew made a major contribution to the book by offering me a copy of his scrupulous report about Joe's final days. David Boufford was kind and forthcoming in response to my queries. I am particularly grateful to Guinevere Wogan for providing information about the scientific activities of the expedition.

These distinguished herpetologists made major contributions about their own research: Bryan Fry, Tim Halliday, Ulrich Kuch, Malcolm McCallum, Ali Rabatsky, Jesús Rivas, and Romulus Whitaker. Special thanks to James B. Murphy for permission to draw upon his research into the amazing life and career of Grace Wiley.

I am grateful to Alan Rabinowitz for his cooperation and candor.

At the California Academy of Sciences, Stan Blum kindly took time to share his memories of his friend with me, as did Peter Fritsch, who also provided me with documents relating to his collaboration with Joe Slowinski in Yunnan. Warm thanks to the staff of the Academy library: Anne Malley, chief librarian at the time I was researching this book, gave me free run of the library's archive of Joe Slowinski's private papers; her assistants Karren Ellsbernd and Larry Currie were also very accommodating, ferreting out historical and scientific texts for me even after I returned home to Bali.

In Burma, many people aided me on my visit to the country in 2005. I am especially grateful to U Khin Maung Zaw, director of the Nature and Wildlife Conservation Division; his assistant, Tin Tun; Khin Ma

Ma Thwin, secretary of the Biodiversity and Nature Conservation Association; and Hla Tun and Pocho Soe Lwin, members of Joe Slowinski's final expedition. Thanks also to Tin Htut, Yosep Kokae, Joseph Tawng Wa, and U Phungriq Min. (The assistance offered me by these and others in Burma was confined entirely to scientific matters; at no time did we discuss the internal affairs of their country.)

At the United States Department of State, I wish to thank John Haynes, Elizabeth Jordan, and Karl Wycoff, who were posted at the American Embassy in Burma at the time of Joe Slowinski's death, and to Mary Ellen Countryman, who assisted my research in Burma when I visited there in 2005.

Brian Armstrong and Nancy Donnelly at the National Geographic Society were exceptionally welcoming and helpful, by sharing their personal reminiscences and allowing me to view their archive of outtakes from the films they made with Joe Slowinski in Burma.

I am very grateful to Ashleigh Smythe and Sandy Scoggin for their candor and trust in talking about their experiences with Joe Slowinski. Thanks also to Craig Maughan, Stanley Rasmussen, and Sean Windsor for their insights into Joe's early years in Missouri and Kansas. Jennifer Durand's account of Joe Slowinski's role as a mentor to her son made an important contribution to my evolving understanding of him, even if it didn't form a part of the final text.

I drew on the published research of many scientists and journalists, in nearly every case identified in the text of the book. Articles by science reporters Steve Grenard, Jim Herron Zamora, and Carl Zimmer, and particularly David Perlstein's excellent features about Joe Slowinski in the *San Francisco Chronicle*, were an important resource. There is a surprising paucity of good (or even mediocre) books about contemporary herpetology; but I consulted two masterworks, quite different in scope and intention, almost daily while I was writing this book: Harry Greene's *Snakes: The Evolution of Mystery in Nature*, and *Herpetology: An Introductory Biology of Amphibians and Reptiles* by George R. Zug, Laurie J. Vitt, and Janalee P. Caldwell. For information about the history of the Kachin and their conversion to Christianity, I relied upon *Among*

Insurgents, Shelby Tucker's engrossing account of a trek through Burma in the late 1980s. The quotation from the Book of Genesis is taken from the Revised Standard Version.

While I have been the beneficiary of an extraordinary amount of expert assistance, I alone am responsible for any errors or omissions in this book.

In addition to her contributions to the text, Rachel Slowinski also played a key role in gathering and preparing for print the photographs published in this book. Her and her parents' archive of photographs was the source for most of the first photo section, devoted to Joe's early years: I am grateful to the family for this important contribution. The second photo section, illustrating Joe's career in Burma, draws largely upon the magnificent photography of Dong Lin, who graciously allowed me to publish his pictures of his friend. Thanks to Gregory Farrington, executive director of the California Academy of Sciences, for granting me permission to publish them.

I am also grateful to Jens Vindum for giving me invaluable access to the photographic archive of the Myanmar Herpetological Survey when I was in Yangon. Other friends and colleagues of Joe's who generously contributed photographs for use in this book are Brian Armstrong, Brian Crother, Nancy Donnelly, Stanley Rasmussen, Alan Rebertus, and Ashleigh Smythe. Thanks also to Bruce Granquist, my neighbor in Bali, for creating the beautiful endpaper maps and snake vignettes for the chapter headings.

Friends in Indonesia provided support and inspiration to me while I was writing the book: Jerry Chamberland, Nancy-Amelia Collins, Richard Howells, Karin Johnson, Henry M. Rector Jr., and Dean Tolhurst. In Bali, Made Wijaya listened to my complaints about the bumps in the road with unfailing good cheer. John McGlynn, in Jakarta, has been my mainstay—and best reader—since I came to Indonesia to live, in 1999. Mark Livingston, in Berkeley, and Craig Seligman and Silvana Nova, in Brooklyn, helped me even more with their creative thinking and learned conversation than with their splendid hospitality while I was carrying out research for the book. Old Burma hand Russell Ciochon introduced me to Southeast Asia and the maddening joys of doing science there many years ago, and has been a source of support ever since.

As always, I wish to thank my agent, Katinka Matson, for her loyalty and encouragement. Will Schwalbe, my editor, challenged me to think in new and better ways: I am enormously grateful to him for his unflagging belief in this book. Thanks also to Brendan Duffy, who capably served as the point man in the production of the book. Alice Truax applied her expert eye and pitch-perfect ear to the manuscript at a late stage, and suggested many excellent improvements.

Finally, as ever my deepest gratitude goes to my partner, Rendy Bugis, for his steadfast kindness and patience.

Index

Note: **Bold** page numbers refer to species descriptions.